Operational Control of Coagulation and Filtration Processes

AWWA MANUAL M37

Third Edition

American Water Works Association

Manual of Water Supply Practices — M37, Third Edition

Operational Control of Coagulation and Filtration Processes

AWWA Publications Manager: Gay Porter De Nileon
Project Manager: Martha Ripley Gray
Cover Art: Cheryl Armstrong
Production: Darice Zimmermann, Zimm Services; George Zirfas, CDA
Manuals Specialists: Molly Beach, Beth Behner

Library of Congress Cataloging-in-Publication Data

Operational control of coagulation and filtration processes. -- 3rd ed.
    p. cm. -- (AWWA manual ; M37)
    Includes bibliographical references and index.
    ISBN 978-1-58321-801-3
    1. Water--Purification--Coagulation. 2. Water--Purification–Disinfection. I. American Water Works Association.

 TD455.O65 2010
 628.1'64--dc22   2010025238

Printed in the United States of America
American Water Works Association
6666 West Quincy Avenue
Denver, CO 80235-3098

Printed on recycled paper

Contents

Figures

This page intentionally blank.

Tables

This page intentionally blank.

Acknowledgments

The first edition of AWWA Manual M37 (1992) was prepared by the Coagulation and Filtration Committee of the AWWA Water Quality Division under the direction of David A. Cornwell, who served as overall coordinator and technical editor.

The second edition of the manual (2000) was also prepared by the Coagulation and Filtration Committee of the AWWA Water Quality Division under the direction of David J. Hiltebrand, with special assistance from Peter Pommerenk in reviewing the manual for content and consistency. Matt Alvarez and Gary Schafran provided additional reviews and recommendations.

The third edition of the manual was also prepared by the Coagulation and Filtration Committee of the AWWA Water Quality Division under the direction of Cory Johnson and Elizabeth Pyles, with special assistance from Gary Logsdon, who served as the technical editor. Gary's assistance and familiarity with previous M37 editions were invaluable to the completion of the third revision.

Authors of M37 include:

Chapter 1:   Kwok-Keung (Amos) Au, Greeley and Hansen, Chicago, Ill.; Scott M. Alpert, Hazen and Sawyer, Charlotte, N.C.; David J. Pernitsky, CH2M HILL, Calgary, Alta.

Chapter 2:   Susan Teefy, Water Quality & Treatment Solutions Inc., Castro Valley, Calif.; James Farmerie, ITT Water & Wastewater, Zelienople, Pa.; Elizabeth Pyles, Orica Watercare Inc., Dry Ridge, Ky.

Chapter 3:   Robert Bryant, Chemtrac Systems Inc., Norcross, Ga.; Michael Sadar, Hach Company, Loveland, Colo.; David J. Pernitsky, CH2M HILL, Calgary, Alta.

Chapter 4:   George Budd, Black & Veatch, Harborton, Va.; James Farmerie, ITT Water & Wastewater Products, Zelienople, Pa.; Paul Hargette, Black & Veatch, Greenville, S.C.

Chapter 5:   Kevin Castro, GHD Inc., Cazenovia, N.Y.; Rasheed Ahmad, Department of Watershed Management, City of Atlanta, Atlanta, Ga.

Chapter 6:   Orren Schneider, American Water, Voorhees, N.J.; James Farmerie, ITT Water & Wastewater, Zelienople, Pa.; Gary Logsdon, Consultant, Lake Ann, Mich.

Chapter 7:   A compilation of case studies with authorship of each case study listed

Case Study 1:   George Budd, Black & Veatch, Harborton, Va.; George Duval, Chesterfield County, Va., Midlothian, Va.

Case Study 2:   George Budd, Black & Veatch, Harborton, Va.; Paul Hargette, Black & Veatch, Greenville, S.C.

Case Study 3:   George Budd, Black & Veatch, Harborton, Va.; Paul Hargette, Black & Veatch, Greenville, S.C.; Bill Brewer, Winston-Salem/Forsyth County, City/County Utilities, Winston-Salem, N.C.

Case Study 4: Tom Elford, City of Calgary, Calgary, Alta.; David J. Pernitsky, CH2M
HILL, Calgary, Alta.
Case Study 5: David Teasdale, City of Kamloops, Kamloops, B.C.
Case Study 6: Michael Sadar, Hach Company, Loveland, Colo.
Case Study 7: Michael Sadar, Hach Company, Loveland, Colo.
Case Study 8: Robert D. Cummings, Clackamas River Water, Clackamas, Ore.
Case Study 9: Tim McAleer, Palm Beach County Water Utilities, Palm Beach, Fla.;
Jose Gonzalez, PVS Technologies, Detroit, Mich.

This manual was approved by the AWWA Coagulation and Filtration Committee.
Members of the committee at the time of approval of this third edition were as follows:

R. Ahmad, City of Atlanta, Department of Watershed Management, Alpharetta, Ga.
S. Alpert, HDR Engineering Inc., Charlotte, N.C.
A. Au, Greeley and Hansen, Chicago, Ill.
R. Brown, EE&T, Newport News, Va.
B. Bryant, Chemtrac, Norcross, Ga.
G. Budd, Black & Veatch, Harborton, Va.
K. Castro, GHD, Cazenovia, N.Y.
S. Clark, HDR Engineering Inc., Denver, Colo.
K. Comstock, Brown & Caldwell, Atlanta, Ga.
S. Crawford, CDM, Dallas, Texas
J. Farmerie, ITT Leopold, Zelienople, Pa.
T. Getting, ITT Leopold, Zelienople, Pa.
J. Gonzales, PVS Technologies, South New Berlin, N.Y.
S. Hardy, Hazen & Sawyer, Atlanta, Ga.
P. Hargette, Black & Veatch, Greenville, S.C.
E. Harrington, AWWA Staff Advisor, Denver, Colo.
C. Johnson, CH2M HILL, Orlando, Fla.
W. O'Neil, CDM, Carlsbad, Calif.
D. Pernitsky, CH2M HILL, Calgary, Alta.
J. Pressman, USEPA, Cincinnati, Ohio
E. Pyles, Orica Watercare Inc., Dry Ridge, Ky.
M. Sadar, Hach Company, Loveland. Colo.
O. Schneider, American Water, Voorhees, N.J.
S. Teefy, Water Quality & Treatment Solutions, Castro Valley, Calif.

Introduction

The first successful practice of water filtration in the United States involved use of slow sand filters in which raw water was applied directly to large sand beds, but these filters were not suitable for treatment of muddy river waters like those found in the Ohio, Mississippi, and Missouri River valleys and their tributaries. In the 1890s and very early 1900s, George Fuller's filtration tests in Louisville and Cincinnati and Alan Hazen's testing program in Pittsburgh showed that turbid waters could be treated successfully by addition of coagulant chemical, clarification, and rapid sand filtration. The capability of a process train consisting of coagulation, mixing, flocculation, sedimentation, and rapid sand filtration to treat raw water having a wide range of turbidity resulted in widespread acceptance of this process train, which came to be called conventional treatment in the United States. Adoption of conventional treatment by a large number of water systems and of chlorination by even more water systems resulted in a very large decrease in the number of cases and number of deaths caused by typhoid fever in the early decades of the twentieth century.

Prior to World War II the focus on water treatment was on disinfecting water and providing clear water to drink. Coagulation and filtration had been shown to remove a substantial fraction of bacteria from water, and combined with chlorination, conventional treatment provided a double barrier against passage of pathogenic bacteria into drinking water. With the realization that viruses also could be transmitted by drinking water, the microbiological challenge broadened. Conventional treatment was found to be capable of removal of polioviruses in the 1960s, and in the 1980s and 1990s removal of protozoan cysts was shown to be within the capabilities of coagulation and filtration when these processes are managed properly. Results of studies on removal of asbestos fibers by coagulation and filtration proved that this process could remove both microbes and inorganic particles in a very wide range of sizes, from considerably less than 1 μm to tens of μm.

Regulatory requirements related to turbidity of filtered water have become more stringent over the decades, but regardless of the regulatory requirement, the drinking water industry has been able to look to some water systems that set their own goals for filtered water turbidity that were considerably more stringent than those set by regulators. This continues to be the case, as at some filtration plants the operating goal is to produce filtered water turbidity of 0.1 ntu or lower. The Partnership for Safe Water encourages the approach of continually striving to improve filtered water quality. Research for removal of viruses, bacteria, protozoan cysts, and asbestos fibers supports the concept that attaining very low filtered water turbidity is an effective means of consistently attaining the best removal of particulate contaminants. Employing proper coagulation chemistry is fundamental to successful filtration for controlling particulate contaminants.

In addition to playing such an important role in removal of particles in granular media filtration, coagulation also has had other important applications, and new ones are being identified. For precipitative lime softening plants that do not soften at a high pH and remove magnesium, the calcium carbonate crystals that are precipitated in the softening process carry a negative charge, and use of a positively charged coagulant or polymer aids in effective clarification and filtration. When surface waters are softened in this manner, use of a coagulant is required by the Surface Water Treatment Rule (SWTR). Depending on the nature of natural organic matter (NOM) found in water, chemical coagulation can be effective for removing a substantial fraction of the NOM. Rapid oxidation of reduced iron and arsenic results in floc formation with sorbed arsenic on the iron floc, and this can be an effective approach to arsenic removal. Coagulation has also proven to be useful in pretreatment of some waters for membrane filtration.

With the discovery in the 1970s of the formation of trihalomethanes (THMs) in drinking water because of chlorination, an additional purpose beyond control of turbidity was found for coagulation and filtration. Early studies of THMs indicated that three control strategies could be pursued:

- Change to a disinfectant that did not form trihalomethanes

- Remove NOM that reacts with chlorine to form THMs

- After THMs are formed, treat water to remove them

Treating water to remove THMs generally was not practical, so much of the effort to control these compounds focused on changing to a disinfectant that would not form THMs and removing NOM prior to chlorination. Removing the NOM by applying coagulation and clarification in a more effective manner, combined with delaying the introduction of chlorine into water until after clarification was completed, was shown to be an economical means of lowering the concentration of THMs in some waters. Thus the benefits of effective coagulation and clarification were extended beyond removal of turbidity-causing particles and removal of microorganisms.

With the passage of increasingly stringent regulations on the concentration of disinfection by-products (DBPs) in drinking water, removal of NOM has become a regulation-driven goal for many water utilities that depend upon surface water sources and even for some that treat groundwater. For many utilities, meeting both surface water treatment regulatory requirements for filtered water turbidity and the requirements for DBPs can be challenging. NOM often provides an important contribution to the negative surface charges found on both organic and mineral particles, so the nature of NOM and its concentration in water can have a strong influence on the type and dosage of coagulant needed for optimizing coagulation, clarification, and filtration.

More recently, as the merits of the microfiltration and ultrafiltration processes have been recognized and costs of the process equipment have become more affordable, ways have been sought to extend the use of these processes that simply strain particulate matter out of water but do not remove dissolved constituents. Again chemical coagulation has been recognized as a process that could pretreat water prior to membrane filtration and thus extend the range of water quality that can be treated this way. Coagulation for removal of NOM, when the NOM is susceptible to removal by this technique, has proven to be an excellent pretreatment for use in conjunction with membrane filtration to control both particulate contaminants and organic matter that can serve as the precursor to DBPs.

Coagulation is important for many goals of water treatment, so chapter 1, "Particle and Natural Organic Matter Removal in Drinking Water Treatment," deals extensively with this topic. The influence of NOM on coagulation is explained, along with the role of pH and solubility of metal coagulants.

Determining the appropriate chemical conditions, coagulant, and sometimes polymer dosages for coagulation and flocculation is a necessary step at plants where coagulation is practiced. Chapter 2, "Jar Testing," presents extensive information on procedures for using jar tests to determine the conditions needed for successful treatment full-scale.

Chapter 3, "Online Sensors for Monitoring and Controlling Coagulation and Filtration," was prepared because numerous measurements, both chemical and physical, are needed in water treatment plants on a daily basis. This is especially so for plants treating surface water, as the Surface Water Treatment Rule and its subsequent modifications have imposed a significant regulatory requirement for monitoring. In addition, the quality of some surface waters can change substantially over one working shift, or even more rapidly. To maintain the careful process control over chemical coagulation and subsequent treatment steps, online monitoring devices are available and can greatly reduce the burden on operators who would otherwise have to perform many analytical procedures manually. With the convenience of online monitoring, however, comes the necessity to maintain an excellent quality control program so the operations staff and management know that they can have confidence in the results being obtained from the online instruments. Online monitoring can be especially helpful in plants that employ high-rate clarification processes or direct filtration, as the residence time in such plants is often much shorter than the residence time in conventional water filtration plants. For continuing effective water treatment at plants with shorter residence times, online monitoring is needed to alert operators to any adverse changes in raw or treated water quality so prompt corrective action can be taken or so operators can verify that management of chemical feeds by online instrumentation has been done correctly and treated water quality goals continue to be met.

Treatment of coagulated water to create floc growth and to remove suspended solids by clarification is discussed in chapter 4, "Flocculation and Clarification Processes." Information is presented on a wide range of traditional and newer clarification processes in this chapter.

Even as new applications are found for coagulation, the main purpose for which it is used is to condition water for clarification followed by filtration in rapid rate granular media filters. Even if coagulation is done properly, mismanagement of granular media filters still can result in impaired filtered water quality. In order to optimize filter performance, operators need to understand how to manage tasks such as filter backwashing, returning filters to service, and imposing rate increases on filters. These topics are addressed in chapter 5, "Filtration," along with a discussion of particle removal mechanisms in granular media filters and biological filtration.

Chapter 6, "Pilot Testing for Process Evaluation and Control," presents information for those who are considering undertaking pilot filter column or pilot plant water treatment studies to evaluate process modifications or new treatment approaches on an existing water source or to explore treatment options for a new source of water. This chapter also presents a description of the use of pilot filter columns as an online process control tool for assessing the adequacy of coagulation in the full-scale plant.

Practical examples related to information presented in earlier chapters may be found in chapter 7, "Case Studies." When the topic of a case study in chapter 7 is relevant to text in an earlier chapter, it is mentioned in the earlier chapter.

Even with all of the instrumentation, mechanization, and computerization of operations in water treatment plants, the human factor remains vitally important. In a 1989 Awwa Research Foundation (now Water Research Foundation) report entitled *Design and Operation Guidelines for Optimization of the High-Rate Filtration Process: Plant Survey Results,* John L. Cleasby and his co-authors emphasized the human factor. Among their conclusions about the key factors contributing to successful high-rate filtration resulting in low-turbidity finished water were the following:

1. Management must adopt a low turbidity goal, convince the operators that this is a serious goal to be met, and budget adequate funds for whatever chemical dosages are required to achieve the goal. Chemical pretreatment prior to filtration is more critical to success than the physical facilities at the plant. However, good physical facilities may make achievement of the goal easier and more economical. ...

7. Good operator training and the building of operator pride in quality of the treated water are important steps in producing the best filtered water. Some plants utilize 12 hour operating shifts to give more continuity to plant operation, and a short period of shift overlap to provide for intershift communication related to the current treatment strategy.

The advice given by Cleasby and his co-authors is sound. Water treatment plant operators work to produce the drinking water that is supplied to them and their relatives, friends, neighbors, and community in general. The health protection of all in the community is a function of those who operate and oversee water treatment plants. Over the last 100 years or more, the drinking water industry in the United States has made great progress in diminishing health risks related to drinking water. The incidence of waterborne disease is much, much lower than it was in the 1890s, thanks to the many improvements in water treatment that have been implemented in the United States. An important purpose of this manual is to promote the continued improvement in drinking water treatment in future years by providing current information on this topic.

REFERENCES

Cleasby, J.L., A.H. Dharmarajah, G.L. Sindt, and E.R. Baumann. 1989. *Design and Operation Guidelines for Optimization of the High-Rate Filtration Process: Plant Survey Results*. Denver, Colo.: Awwa Research Foundation and AWWA.

Chapter **1**

Particle and Natural Organic Matter Removal in Drinking Water

Kwok-Keung (Amos) Au, Scott M. Alpert, and David J. Pernitsky

INTRODUCTION

One of the most basic processes in the treatment of raw source waters to meet drinking water standards is the solid/liquid separation process to remove particulate material. Particulate material originating in raw water or contributed by addition of treatment chemicals is physically separated from source water during drinking water treatment by clarification and filtration processes. These processes target not only removal of particulate material itself but also contaminants that are associated with the particulate material. Clays, sands, colloids, and so on all may comprise typical particulates to be removed; however, removal of other particle classes, such as microorganisms and particulate forms of natural organic matter (NOM), is beneficial for efficient treatment. Further, other contaminants (e.g., arsenic, iron, manganese, or dissolved NOM) may be associated with particulate matter via coprecipitation, sorption, or other physico-chemical mechanisms. Disinfection by-products (DBPs) have been a primary driver for specific focus on NOM removal. In fact, although much research has been devoted to the coagulation of inorganic particles, coagulant dosages for many surface waters are controlled by the NOM concentration rather than by turbidity. During coagulation, dissolved-phase NOM is converted into a solid phase, allowing removal in subsequent clarification/filtration processes. Finally, chemical and/or physical disinfection is also dependent on effective removal of particulate matter that may shield microorganisms from disinfectant contact and/or reduce the effectiveness of disinfection chemicals.

This chapter provides an overview on the removal of particles and NOM by coagulation and filtration processes. These fundamentals serve as a basis for compliance with multiple treatment objectives and prepare the reader for additional detail introduced later in this manual. Specifically, the following are included in this chapter:

- A review of particles and NOM, including the characteristics of these constituents important in their removal

- A discussion of particle surface charge and coagulant chemistry

- An overview of the physical and chemical aspects of coagulation and filtration processes

- A brief discussion of management of multiple processes for effective treatment, including the multiple barrier approach, process control, and membrane filters

PARTICLES

Particles are ubiquitous in all natural waters. Their origins, compositions, and concentrations vary widely. They can be contaminants (as defined here as substances of natural, anthropogenic, or microbiological origin that may be harmful to the public health, adversely affect water quality, and/or affect the aesthetic properties of the finished water) or associated with contaminants and therefore need to be removed. Of the many ways that particles can be characterized, size and surface properties are two of the most important keys related to removal by coagulation and filtration processes. Further, different techniques exist that can be used to quantify or characterize particles. In this section, particle characteristics, quantification, and water quality are discussed.

Origin and Composition of Particles

Based on their underlying composition, particles can be considered as organic, inorganic, or biological (a subset of organic). Particles are introduced into natural waters (i.e., our water supplies) either through natural processes or as a result of anthropogenic (human) activities. An overview of the three major compositions of particles is provided below:

- The majority of organic particles in natural water are the result of degradation of plant and animal materials. These organic constituents may be classified as natural organic matter. However, NOM in natural water consists of more than suspended particles; it also includes dissolved NOM molecules. A separate section of this chapter focuses on NOM and its removal because of its increasing importance to water utilities.

- The majority of inorganic particles in natural water are mineral particles. Most of these particles are derived from the natural weathering of minerals. Examples include clays, iron oxides, aluminum oxides, and calcites. Inorganic particles often enter source water by means of erosion and runoff.

- Biological particles include microorganisms such as viruses, bacteria, and protozoa. These microorganisms enter the water through direct discharge of wastewater, runoff from the watershed, or animal excrement, and some may grow and prosper in the water body. Microorganisms may also be attached to suspended particulate matter. Another type of biological particle is algae, which use mineral nutrients (nitrogen and phosphorus) and photosynthesis to grow.

Particles may result from specific human activity, for example, discharge of municipal or industrial wastewater effluent into a source water. Runoff from land-disturbing activities and property development also introduce particulate matter, into water sources. The composition of these particles is case-specific. The influx of particles into a water body may be driven by natural events such as runoff from snowmelt and precipitation. Depending on the nature of the watershed, particles coming from these events may contain organic, inorganic, and/or biological matter.

In addition to natural particles that occur in water, particles are often added or created in water treatment processes to remove other particles and contaminants. Metal-based coagulants added to water to destabilize particles (as discussed in a later section) may also precipitate as metal hydroxides. These precipitates can then flocculate with particles from the source water and be removed by the solid/liquid separation processes in treatment plants. Bentonite may be added to low turbidity waters to enhance the contact opportunity between particles so that larger and denser flocs are formed for better removal in the settling process. Chlorine may be added to water to oxidize soluble iron (II) species into insoluble iron (III) particles so that these particles can be removed by settling and filtration processes.

Because of their reactivity, pure inorganic or organic particles seldom exist in natural water; that is, all inorganic particles have some kind of organic properties and vice versa. Most inorganic particles have an affinity to and can sorb organic chemicals such as synthetic organic chemicals onto their surfaces. Most inorganic particles in water also react with NOM and form an organic coating on their surfaces. This organic coating plays a significant role on the surface properties of particles, as will be discussed later. Furthermore, NOM, regardless of whether it is in dissolved or particulate form, can chemically bind with many inorganic contaminants such as metals.

The Need to Remove Particles

Particles must be removed from water for both aesthetic and health reasons. The presence of particles may impart color, taste, and/or odor to water, making it less palatable for the customer. More importantly, particles can also be pathogenic or toxic and must be removed to protect public health. Further, particles can shield microbes from disinfectants and reduce the efficiency of the disinfection process. For these reasons, it is essential that coagulation, flocculation, sedimentation, and filtration processes are properly designed and operated to optimize the removal of particles.

Current regulatory requirements for particle removal are based primarily on improved control of microbial pathogens. These requirements are summarized in Table 1-1 and discussed below:

- Removal of particles is regulated indirectly under the Surface Water Treatment Rule (SWTR) and its various revisions. These requirements apply to utilities using surface water or groundwater under the direct influence of surface water.

- Compliance with particle removal criteria is determined by filtered water turbidity, with regulatory limits ranging from 0.3 to 1.0 ntu, depending on the type of filtered water, percentile value, and location of measurement (see Table 1-1).

A utility meeting the turbidity requirements demonstrates that it can consistently provide good removal efficiency of particles and microbes through its coagulation and filtration processes. This, together with properly managed disinfection, ensures that the finished water leaving the treatment plant is of such quality that it minimizes microbial pathogens, has a physical appearance (low turbidity) that is palatable to the consumer, and is safe to drink.

Table 1-1   Regulatory requirements for particle and NOM removals

Constituents	Regulation	Compliance Indicator	Requirements			
Particles	IESWTR[1][3] LT1SWTR[2][3]	Filtered Water Turbidity	CFWT[4]	<0.3 ntu 95% of the time		
				<1 ntu any time		
			IFWT[5]	<0.5 ntu[6]		
				<1.0 ntu[7]		

			Removal Requirements, %			
NOM	Stage 1 D/DBPR[8]	Removal Percentage of Total Organic Carbon (TOC)[9]	Source Water TOC, mg/L	Source Water Alkalinity, mg/L as CaCO$_3$		
				0–60	>60–120	>120
			>2.0–4.0	35.0%	25.0%	15.0%
			>4.0–8.0	45.0%	35.0%	25.0%
			>8.0	50.0%	40.0%	30.0%

Courtesy of Kwok-Keung Au.

(1)   Interim Enhanced Surface Water Treatment Rule
(2)   Long Term 1 Enhanced Surface Water Treatment Rule
(3)   These requirements apply to utilities using surface water or groundwater under the direct influence of surface water
(4)   Combined filtered water turbidity
(5)   Individual filtered water turbidity
(6)   In any two consecutive measurements taken 15 min apart at the end of the first 4 hr of continuous filter operation after backwash
(7)   In any two consecutive measurements taken 15 min apart
(8)   Stage 1 Disinfectants and Disinfection By-products Rule
(9)   These requirements apply to conventional treatment facilities that use surface water or groundwater under the direct influence of surface water

Particle Quantification

Analytical methods related to the number and size of particles in water are turbidity, particle count, and suspended solids concentration measurements. Each method has advantages and limitations and is used to achieve a different objective. These quantification techniques are discussed below, and turbidity and particle counting are described in detail in chapter 3.

- Turbidity measurement is the most widely used method for assessing particles in water. It does not give a quantitative measure of particles in water but instead indicates the relative clarity of water samples by measuring the amount of light scattered by particles in water samples. The result is reported in nephelometric turbidity units (ntu). The turbidity level of a water sample depends on the physical properties (such as concentration, size, and shape) and the optical properties of the particles contained in the sample. Although the actual relationship among these characteristics is very complicated, the result is sufficient to describe turbidity as a composite measurement based on these properties. Turbidity has been used successfully as a regulatory indicator to assess the removal efficiency of particles and microbes by treatment processes. Turbidity measurements can be performed with grab samples or continuous online instruments.

- Particle counts (particle counting) represent the numerical concentration of particles within finite particle diameter ranges. The results are usually reported as cts/mL at different size ranges. Particle counting can be used as a tool to monitor the performance of removal processes. However, because of several limitations in the application of this technique, particle counts are currently not used for regulatory compliance in drinking water. Particle count can be measured by grab samples or continuous online instruments.

- The mass of particles in a water sample can be quantified as the concentration of suspended solids, which is defined as the total mass of particles retained on a glass fiber filter disc through which a measured volume of water sample has been filtered. The result is reported as mg/L. Materials passing through the filter are defined as dissolved solids. Suspended solid measurement is typically not used to assess the removal efficiency of drinking water treatment processes. Instead, because of the exceedingly small mass of suspended solids in filtered drinking water, this analytical method is used by water utilities to estimate the amount of sludge or waste produced from clarification and filtration processes.

Particle Size

The size of particles is an important characteristic affecting their removal in water treatment plants. Particle size may vary by several orders of magnitude. Most inorganic particles have sizes ranging from 0.1 to 5 micrometers (one-millionth of a meter, or μm). Figure 1-1 shows a comparison of sizes that may be encountered in water supplies. Biological particle size is dependent on the classification of the microorganism. Viruses, for example, are the smallest biological particles and have sizes of 3–100 nanometers (one billionth of a meter, or nm). Bacteria are larger than viruses and have sizes from slightly less than 1 μm to over 10 μm. Algae and protozoan cysts are even bigger and have sizes from a few μm to a several hundred μm.

Operators are often familiar with the terms *colloidal/suspended particles* and *suspended/dissolved solids*. These terms are based on particle size and sometimes can be confusing. Colloidal particles are particles with at least one of their dimensions less than about 1 μm or 0.5 μm and generally are not filtered out in the suspended solids test. Dissolved solids contain both colloidal particles and the impurities that are in dissolved form. By definition, colloidal particles do not include constituents that are in true dissolved or molecular form, which typically have sizes of less than 1 nm.

Particle size is important in water treatment because it is one of the key factors in determining the settling characteristics of the particle. For example, the settling velocity of a particle is directly proportional to the square of its diameter. Natural particles in the colloidal size range do not settle quickly enough to be removed in sedimentation basins, so they must be agglomerated together into larger particles, i.e., floc. The size of the floc particle is important for effective settling. Particles passing through the sedimentation process (and flocculated particles in direct filtration treatment plants) may be removed in the filtration process. Again, the size of the particle determines whether the particle will be removed in the top layer of the filter or will penetrate deeper into the filter bed.

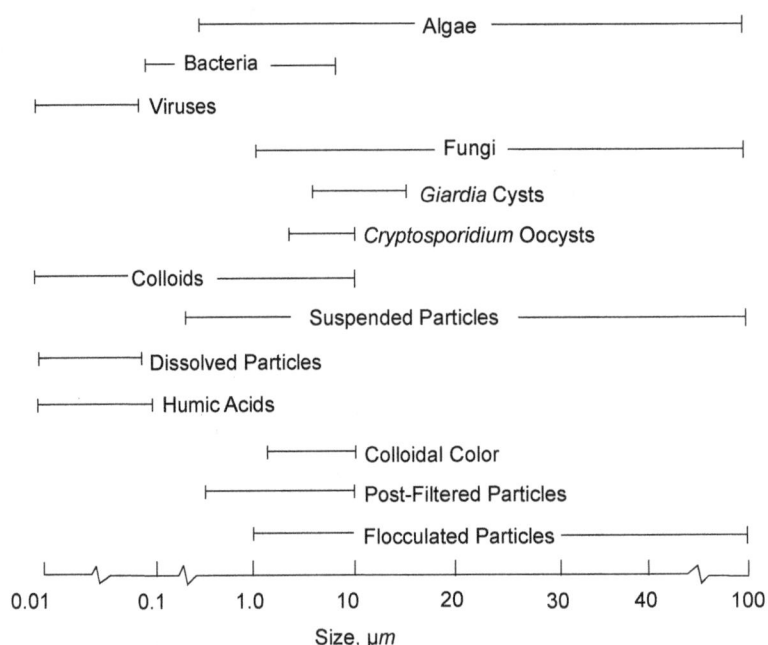

Source: McTigue and Cornwell 1988.

Figure 1-1   Particulates present in source and finished water

NATURAL ORGANIC MATTER

Origin and Composition of NOM

Natural organic matter, or NOM, is a complex mixture of natural constituents. The primary sources of NOM are the degradation of vegetation in the watershed area and the growth and decomposition of aquatic organisms such as algae and weeds within the water body. NOM is generally classified into two components: humic substances (HSs) and nonhumic substances (nHSs). HSs are usually the major components of NOM in water with humic acids (HAs) and fulvic acids (FAs) as the major fractions. The major fractions of nHSs are proteins, polysaccharides, and carboxylic acids.

Both humic acids and fulvic acids represent a broad class of heterogeneous organic materials. Because of their complexity and heterogeneity, they cannot be well characterized in terms of a specific chemical structure. The most important properties of these substances are molecular weight, functional groups, and charge behavior. HAs and FAs are macromolecules with molecular weights of several hundred or higher. HAs and FAs carry weakly acidic functional groups such as carboxylic and phenolic groups. Dissociation of these functional groups induces negative charges of HS. The macromolecular nature and charge behavior of HAs and FAs play a significant role in increasing the colloidal stability of particles that bind with NOM and in the removal of NOM from water. The presence of NOM in water will induce an additional demand for coagulant dose. Two mechanisms are usually considered for coagulation of NOM. The first is the neutralization of negative charges of NOM followed by precipitation of the NOM. The second is adsorption of NOM onto precipitates formed from coagulants.

The Need to Remove NOM

Although NOM itself does not impose a direct health threat, removal of NOM from water is becoming more important for aesthetic, health, and operational reasons. The initial driving force for removing NOM was its adverse effect on aesthetic water quality because the color, taste, or odor caused by NOM can make water much less palatable for the consumer. For example, the presence of NOM molecules can make water appear yellow or brown. More importantly, however, NOM is now known to be the major precursor for many disinfection by-products (DBPs). The chemical compounds known as DBPs are formed when an oxidant such as chlorine is added to water that contains organic matter. Removal of NOM to reduce the formation of DBPs has become a major focus for water utilities in the past two decades. A coagulation process that is optimized for both particle removal and NOM reduction is known as enhanced coagulation (EC).

NOM also has adverse impacts on the operation of other treatment processes. By reacting with chemical disinfectants to form DBPs, NOM induces a disinfectant demand. Some NOM molecules contain chemical functional groups that can absorb light in the ultraviolet (UV) range and thus reduce the efficiency of UV disinfection facilities. NOM can foul membranes, reducing flux or increasing operating pressure. Short-chain molecules of NOM (e.g., as a result of oxidation) can serve as a food source for microbial growth in filter beds or the distribution system.

Current regulatory requirements for particle removal and NOM reduction are based on a balance between adequate inactivation of microbial pathogens and minimizing the production of DBPs. These requirements were summarized in Table 1-1 earlier in this chapter and are discussed below:

- NOM is regulated under the Disinfectants and Disinfection By-products Rule (D/DBPR), with reduction related to requirements for minimum removal percentages of TOC between the source and finished water. These removal levels range from 15 pecent to 50 percent and are based on TOC levels in the source water and the source water alkalinity (Table 1-1).

- Compliance with NOM removal requirements typically indicates that a utility is effectively using its coagulation and filtration processes to significantly reduce the concentrations of DBP precursors and the formation of DBPs.

- Alternative compliance criteria for NOM removal exist (although not shown Table 1-1). These criteria are designed to provide flexibility to water utilities that use source water either with a proven low potential to form DBPs (indicated by low concentrations of TOC, specific ultraviolet absorbance [SUVA], and DBPs) or that have source waters that are not amenable to significant TOC removal as indicated by site-specific bench-scale jar test results.

The key to meeting the challenges set by new regulations will often be maximizing the removal of NOM while ensuring adequate microbial control by both particle removal and disinfection. For many surface waters, coagulant dosages are controlled by NOM concentration rather than by turbidity (Edzwald and Van Benschoten 1990, Pernitsky and Edzwald 2006). NOM can be removed by coagulation through complexation with positively charged coagulant species forming Al-NOM precipitates or by the adsorption of NOM onto the surface of floc particles, allowing removal in subsequent solids separation processes. The charge density of these NOM functional groups is typically 10 to 100 times greater than the charge density of inorganic particles. For example, a water with 10 mg/L of clay turbidity having a negative charge of 0.5 µeq/mg will have a positive charge demand of 5 µeq/L. In contrast, a water containing only 3 mg/L dissolved organic carbon (DOC) with a negative charge of 10 µeq/mg will have a positive charge demand of 30 µeq/L, which is six times that of the 10 mg/L clay turbidity example (Edzwald and Van Benschoten 1990). The conversion of dissolved NOM to a solid phase happens quickly and is complete prior to the clarification and filtration process.

As mentioned previously, NOM is a mixture of various organic compounds that are present in water as a result of decay of vegetation, runoff from organic soils, and biological activity. As such, NOM from different water sources will have different characteristics. The concept of specific UV absorbance has been developed as an operational indicator of the nature of NOM and the effectiveness of coagulation for removing NOM, DOC, and DBP precursors (Edzwald and Van Benschoten 1990, Edzwald and Tobiason 1999). SUVA values offer a simple characterization of the nature of the NOM, based on measurements of the UV absorbance at 254 nm and DOC. SUVA is defined as the UV absorbance of a water sample normalized with respect to the DOC concentration. SUVA is normally calculated on samples of raw water prior to the addition of any treatment chemicals. Samples must be filtered in the lab to remove turbidity interferences as described in Standard Method 5910 (APHA et al. 2005). It is expressed in units of m^{-1} of absorbance per mg/L of DOC, or L/mg C $\cdot$ m^{-1} and also expressed using the units notation of L/mg-m.

$$\text{SUVA (L/mg C} \cdot m^{-1}) = [\text{UV}_{254}(cm^{-1}) \times 100 \text{ (cm/m)}] / \text{DOC (mg/L)}$$

Guidelines for the interpretation of SUVA values are presented in Table 1-2. For supplies with low SUVA (2 or lower), DOC will not control coagulant dosage. For water supplies with SUVA greater than 2, the amount of NOM typically exerts a greater coagulant demand than the amount of particles. For these waters, the required coagulant dosage increases with increasing DOC concentration.

Optimizing treatment processes to remove NOM and particles and also to control microbiological contaminants may involve setting coagulant doses to achieve NOM removal, reducing coagulation pH, improving mixing conditions, and/or using alternative approaches for oxidation of NOM and inactivation of microorganisms. Water treatment operators need effective tools for quick and accurate assessment of treatment performance and evaluations of alternatives. These tools are described in chapters 2 and 3 of this manual.

Table 1-2   Guidelines on the nature of NOM and expected DOC removals

SUVA	Composition	Coagulation	DOC Removals
<2	Mostly nonhumics; low hydrophobicity, low molecular weight	NOM has little influence Poor DOC removals	<25% for alum Potentially higher removals for ferric
2–4	Mixture of aquatic humics and other NOM; mixture of hydrophobic and hydrophilic NOM; mixture of molecular weights	NOM influences DOC removals should be fair to good for these categories	25–50% for alum Potentially higher removals for ferric
>4	Mostly aquatic humics; high hydrophobicity, high molecular weight	NOM controls Good DOC removals	>50% for alum Potentially higher removals for ferric

Source: Edzwald and Tobiason 1999.

PARTICLE STABILITY AND COAGULANT CHEMISTRY _____

Surface Properties and Colloidal Stability

The surface characteristics of natural particles and their reactions with water and with other solutes in water result in an electrical surface charge being carried by most particles in water. More importantly, the sign of the net charge is usually negative under most conditions in water. The following mechanisms are usually used to explain the charge behavior of particles in water:

Lattice imperfection or isomorphic replacement (intrinsic properties of particles). Lattice imperfection refers to the replacement of atoms in the crystalline lattice (structure) by atoms with different valances. This mechanism is often used to explain the charge behavior of many clay minerals. Clay has a layered structure of silica (SiO_2). During the formation of the structure, if an Al^{3+} atom replaces an Si^{4+} atom, a negative charge develops. Similarly, negative charge develops when an Mg^{2+} atom replaces an Al^{3+} atom in an aluminum oxide crystalline lattice (Al_2O_3).

Ionization of particle surface functional groups (reactions with water). Many particle surfaces contain ionizable functional groups. For example, mineral oxide particles contain surface hydroxyl groups. Biological particles may have surface proteins that contain carboxyl and amino groups. In the presence of water molecules, these surface functional groups can accept or donate protons (H^+), depending primarily on water pH. As a result, the surfaces of particles become charged. In this case, the surface charge is strongly pH dependent, being positive at low pH and negative at high pH. Most natural particles have a negative surface charge at the pH of most natural waters.

Reactions between surface functional groups and other solutes in water (reactions with other solutes). Many cations (such as metals) and anions (such as NOM) in water can react with surface functional groups of particles, resulting in the binding of these ions to the particle surfaces. Interactions other than simple electrostatic interactions are often involved in these processes. Typical examples are hydrophobic interaction, hydrogen bonding, ligand exchange, and covalent bonding. As a result, even anions that have negative charges can bind to negatively charged particles. Binding of cations to a particle makes the particle's surface charge more positive, whereas binding of anions to a particle makes the particle's surface charge more negative. An important example is the binding of NOM onto particle surfaces. Recent research indicated that most particles in water carry some kind of NOM coating on their surfaces, one of the major reasons that most particles in water are negatively charged.

Interactions between particles with similar surface charge result in an electrical repulsive force between them, making them more difficult to aggregate. In colloidal chemistry, a *stable* solution or set of particles is defined as one in which most of the particles have similar charges and thus these particles tend not to aggregate or settle. Chemical pretreatment (coagulation) is therefore needed to reduce or eliminate this repulsion and enhance particle removal by sedimentation and filtration processes. This step is termed *destabilization*.

Use of oxidants to change surface properties of particles. Use of oxidants prior to filtration has been shown to benefit filter performance at numerous plants. Reported benefits include a reduction in filtered water turbidity or particle counts or both, a decrease in turbidity peak during the filter ripening period, and a shorter duration for filter ripening. Oxidants generally used for this purpose are free chlorine and ozone. However, other oxidants such as chlorine dioxide and potassium permanganate exhibit similar benefits. A recent study (Becker et al. 2004) indicated that one of the major mechanisms for these benefits is on the effect of oxidation on particle stability. As mentioned previously, most particles in water carry some kind of NOM coating on their surfaces. The NOM coating increases particle stability by making the particle's charge more negative and also extending the particle's negative electric field further away from the particle. Mechanistic studies show that oxidation can detach part of the NOM coating from particle surfaces and thereby reduce particle stability. Typically, very low dosages of oxidant and short contact times are enough for these benefits to occur.

At plants where oxidation is an aid to filtration, interruption of oxidation can cause filter performance to deteriorate. Factors to consider related to using oxidants for improving filter performance are cost of the oxidation process, efficacy of the oxidant, and possible detrimental effects of oxidation products, such as formation of assimilable organic matter and bromate by ozone and formation of chlorinated disinfection by-products by chlorine.

Coagulant Chemistry

Aluminum and iron-based coagulants such as aluminium sulfate (alum), polyaluminum chloride (PACl), and ferric chloride react with water to form charged and dissolved metal-hydroxide species, as well as solid-phase metal-hydroxide precipitates (floc particles). These reactions consume alkalinity in the raw water and reduce the pH. Alum and ferric coagulants are more acidic than PACls and therefore result in greater pH depression after addition. For PACls, alkalinity consumption is related to basicity. Higher-basicity PACls will consume less alkalinity than low- or medium-basicity ones.

The charge on the dissolved coagulant species and the relative amount of floc formed are a function of pH. Therefore, the pH at which coagulation occurs is one of the most important parameters for proper coagulation performance. For alum and PACls, the best coagulation performance is generally seen at pH values that are close to the pH of minimum solubility of the coagulant. This controls dissolved Al residuals, as well as maximizing the presence of floc particles. Acid or base addition is often used after coagulant addition to control pH.

The solubility characteristics of various coagulants, and therefore the pH range at which they are most effective, are important properties of the coagulants. Solubility refers to the maximum concentration of dissolved species that can exist in solution before precipitation. This concentration varies with temperature and pH. The pH of minimum solubility represents the pH at which the concentration of dissolved coagulant species is a minimum. This is important from a treatment perspective, as this pH also corresponds to the point at which the maximum amount of solid floc species is formed.

Table 1-3   Summary of coagulant solubility

Coagulant	Minimum Solubility 20°C		Minimum Solubility 5°C	
	pH	μg/L Al	pH	μg/L Al
Alum	6.0	16	6.2	3
Polyaluminum sulfate (PAS)	6.0	28	6.4	6
PACl low-basicity nonsulfated (LBNS)	6.2	27	6.7	4
PACl medium-basicity sulfated (MBS)	6.3	29	6.5	4
PACl high-basicity nonsulfated (HBNS)	6.4	36	6.8	9
PACl high-basicity sulfated (HBS)	6.4	52	6.9	5
Aluminum chlorohydrate (ACH)	6.7	101	7.6	53
$FeCl_3$	8.7	0.006	—	—

Source: Pernitsky and Edzwald 2003.

The minimum solubility (concentration) and pH of minimum solubility for several common coagulants are shown in Table 1-3. As can be seen in this table, the minimum solubility and pH of minimum solubility differ for the various chemical coagulants. PACls are more soluble and have a higher pH of minimum solubility than alum. Polyaluminum sulfates have solubility characteristics similar to alum. Ferric coagulants are much less soluble than aluminium-based ones. This means that Fe-based coagulants can be used over a much greater pH range without worrying about dissolved metal concentrations in the finished water. The pH of minimum solubility for Fe(III) is near pH 8.8. However, unlike Al-based coagulants, $FeCl_3$ is not an effective water treatment coagulant at its pH of minimum solubility because of the weak positive charge of the $Fe(OH)_2+$ species present at that pH. Ferric coagulants have, however, been used successfully in secondary clarification of lime softening process basin effluent at softening plants that treat surface waters. More effective performance is seen at lower pH, as low as pH 5.5, where more positively charged species are present. A case study, "Conversion From Alum to Ferric Sulfate at the Addison-Evans Water Treatment Plant, Chesterfield County, Va." is presented in chapter 7.

For all of the metal coagulants, it is important to note that the pH of minimum solubility increases as temperature decreases, as shown in Table 1-3. This is especially important in cold climates because of the wide range in raw water temperatures experienced. For example, the pH of minimum solubility for alum increases from 6.0 at 20°C to 6.2 at 5°C. Over that same range, the pH of minimum solubility for high-basicity PACl changes from 6.4 to 6.8.

Particles can be destabilized through the addition of coagulants/flocculants by three major mechanisms (shown in Figure 1-2).

Adsorption and Charge Neutralization

When metal coagulants are added to water, several hydrolysis species are formed. Some of these species are positively charged, depending primarily on water pH. These positively charged species will attach to negatively charged particles and reduce or neutralize the particles' negative charges. This charge neutralization results in a reduction or elimination of the electric repulsion between particles. Cationic polyelectrolytes also can reduce the negative charges and repulsive forces. Note, however, that if the dosage of a cationic polymer is substantially greater than that needed to neutralize the negative charges on particles, then the particles can become positively charged and restabilized, a condition that hinders particle removal.

Enmeshment in a Precipitate (Sweep Floc)

When metal coagulants are added to water, precipitates of metal hydroxide or metal carbonate may form, depending on the dose and water chemistry. Particles can be enmeshed into these amorphous precipitates (coagulant flocs) and subsequently removed by settling and filtration of the flocs. Thus for the sweep floc mechanism, as opposed to the charge neutralization mechanism, the use of dosages of metal coagulants larger than those needed to neutralize the surface charges of particles does not hinder particle removal, because the coagulant will precipitate and can enmesh the particles.

Adsorption and Interparticle Bridging

When high-molecular-weight polymers are added to water, part of the polymeric chains can attach to the surface of one particle with the remaining length of the chains extending into the solution. If these extended chains find other particles with vacant sites not

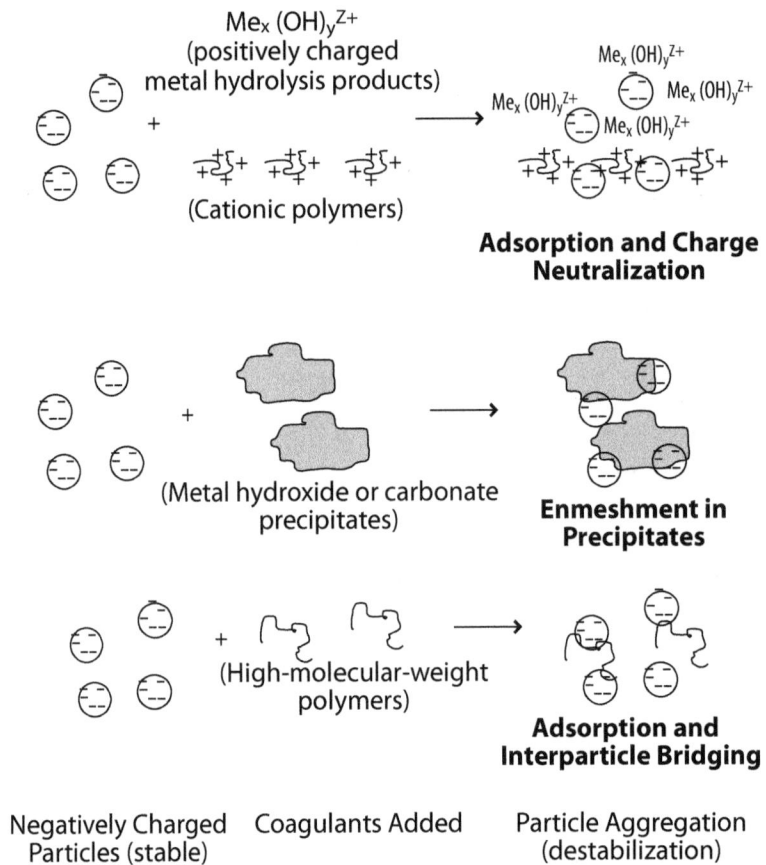

Courtesy of Kwok-Keung Au.

Figure 1-2   Coagulation (destabilization) mechanisms for particulate contaminants

yet attached by other polymeric chains, bridges between particles could form, resulting in particle destabilization and floc formation. Overdose of polymer may result in restabilization because it becomes difficult for the extended polymer molecule to find available vacant sites for adsorption.

Double-layer compression is often cited as the fourth mechanism of coagulation. However, this process is not a dominant mechanism in the chemical coagulation of most raw waters.

PARTICLE AND NOM REMOVAL PROCESSES

The objectives of particle removal and NOM reduction typically cannot be accomplished in a single treatment step. Rather, several plant processes work together to achieve this goal. Specifically, coagulation/mixing, flocculation, sedimentation, and filtration, schematically shown in Figure 1-3, are all interdependent on each other to produce a water of high quality. This section provides a brief overview of each process and describes the mechanisms by which each contributes to particle and NOM removal. These processes are described briefly in this section and in greater detail in chapter 4 (Flocculation and Clarification Processes) and chapter 5 (Filtration).

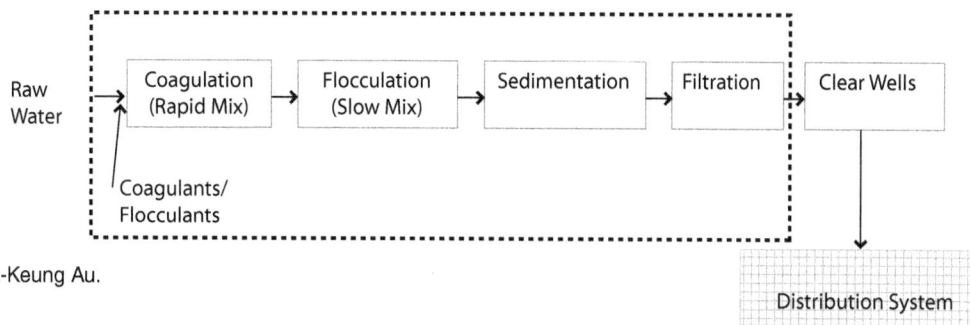

Courtesy of Kwok-Keung Au.

Figure 1-3   Treatment train (from coagulation to filtration)

Mixing/Coagulation

As described previously, many of the particles that occur in raw water supplies have negative electrical charges and are of such size and density that they will not settle easily in the time available in water treatment plant clarification processes. Therefore, positively charged metal coagulants or polymers are used to decrease the extent of the negative surface charge on the particles so that when they come in contact with each other they can stick together and form larger particles (flocs).

Generally, the chemical reactions associated with coagulation occur very quickly and thus the coagulant must be mixed into the raw water as quickly and efficiently as possible. This initial mixing is often called rapid mixing or flash mixing. The hydraulic retention time in a flash mix process ranges from <1 sec to 30 sec. During this time, the coagulant and any other associated chemicals (e.g., pH and alkalinity control) are dispersed throughout the raw water. The dose of chemicals and the required mixing intensities for optimum coagulation can be determined using bench-scale experiments such as jar tests as described in chapter 2. Rapid mixing is discussed in chapter 4. One exception to the very rapid coagulation reaction is the slower action of alum in very cold water (about 5°C or colder). To accommodate the slower reactions, some water utilities use coagulants other than aluminum sulfate for coagulation of such water (Logsdon et al. 2002).

Flocculation

The next treatment process after coagulation typically is flocculation. The main objective of flocculation is to bring together the particle solids created and/or conditioned in the coagulation step, which ultimately changes the size distribution of the particles. Essentially, a large number of small particles are transformed into a smaller number of larger particles. Traditionally, the objective of flocculation has been to produce particles large enough and dense enough to settle in the clarifier (sedimentation basin). Since the 1980s, several plants have replaced conventional sedimentation with filtration without clarification (direct filtration) or dissolved air flotation (DAF). For these types of treatment processes, the goal of flocculation and floc size production is modified since both of these processes work well with flocs that are considerably smaller than the size of flocs needed for sedimentation. For direct filtration and DAF plants, shorter flocculation times are used as compared to the times employed at plants with conventional sedimentation basins.

Sedimentation/Clarification

After the particles have been preconditioned in the coagulation process and brought together into clumps (flocs) in the flocculation process, physical removal of the solids can be accomplished by sedimentation or by flotation. The hydraulic retention time of a conventional sedimentation basin ranges from 2 to 4 hr. Since the late 1960s, studies of the sedimentation process have led to a variety of approaches that can accomplish sedimentation in times considerably shorter than 2 to 4 hr. Clarification, whether by sedimentation or flotation, is necessary for effective filtration of many source waters because of the excessively high load of solids (particulate matter, including floc formed by coagulant chemical) that would be applied to the filters in the absence of a clarification process.

Filtration

After clarification, water is treated by filtration to remove those particles that were not removed in the clarification process. As water treatment was being developed in the United States, experimental work in Louisville, Ky., and Cincinnati, Ohio, in the late 1890s and early 1900s showed that turbid waters could be treated successfully by chemical coagulation, sedimentation, and rapid sand filtration. About the same time, studies at Pittsburgh, Pa., showed the importance of using an adequate dosage of coagulant chemical to attain successful treatment with rapid sand filters. Coagulation, clarification, and rapid sand filtration became known as conventional water treatment, and this process train was shown to significantly reduce both turbidity and bacteria in water. Today, the performance of granular media filters reflects both source water quality and the changes to the source water induced by added chemicals (pretreatment chemistry). Filtration in drinking water treatment is not just a physical straining process (like that in a coffee filter) by the granular media particles. Attachment of the particles to the filter media is the primary form of target constituent removal. Thus, filtration is a physical and chemical process in which the effectiveness of the particle removal is determined by several variables, including:

- Type of filter media (size, depth, material)

- Water chemistry

- Surface chemistry of the particles (as conditioned by coagulation and flocculation)

- Surface chemistry of the filter media.

During filtration, particles must be transported to the surfaces of the filter media, and the particles must attach to the media surface for removal to occur. Both hydrodynamics and chemistry are important determinants of success. Design criteria most often specify filtration rate, media size, and bed depth. Pretreatment (coagulation) chemistry is the most important factor affecting particle removal in granular media filters. Without proper coagulation, efficient particle removal will not occur.

Plant operators have direct control over the coagulation process (chemical selection and dosing), flocculation (mixing energy), filtration (filter run times, backwashing), and, to some extent, flow rates through each of these processes. Thus, operators have the ultimate responsibility to ensure effective particle and NOM removal.

MULTIPLE BARRIER APPROACH

Multiple treatment processes must be incorporated into a water treatment plant to achieve high-quality finished water. Combined with disinfection, clarification and filtration processes provide multiple barriers to the passage of particles, pathogens, and dissolved constituents into the public water supply. This multiple barrier approach was formally established in the Surface Water Treatment Rule for microbial control and removal of contaminants, and specifically referenced the coagulation/filtration processes, primary disinfection as defined by the CT concept, and maintenance of microbial control through the distribution system. The simple combination of coagulation, sedimentation, filtration, and disinfection constituted an early multiple-barrier approach to microbial control in drinking water technology.

As more contaminants, including pathogens other than bacteria, were discovered in raw water supplies, drinking water treatment objectives were expanded. The drinking water processes of coagulation, flocculation, sedimentation, and filtration must now be optimized to meet multiple treatment objectives. Further, new advanced technologies are being developed and may be incorporated within the conventional treatment process train.

One of the more effective methods of meeting new regulatory requirements is through a "systems approach" that recognizes that all unit processes are interrelated, so what impacts one will also impact the others. Analysis under this systems approach reveals that the conventional treatment processes work together to remove unwanted components from the water, including particles, NOM, color, microorganisms, iron, manganese, and objectionable tastes and odors. Coagulation and flocculation create flocs of suspended particles and convert organic and/or inorganic material from the dissolved phase into the particulate phase. These conditioned particles are subsequently removed by either clarification or filtration.

Much of the effort to optimize conventional treatment focuses on coagulation chemistry as the single most important factor affecting treatment plant performance. This principle is based on the fact that if the pretreatment chemistry is wrong, none of the other downstream processes will work well. Managing the coagulation process to remove both particles and NOM is an example of the challenges that may be encountered when it is necessary to adjust pretreatment chemistry to achieve multiple objectives.

PROCESS CONTROL STRATEGIES

Effective process control strategies are based on theory, experience, practical knowledge of the source water, and performance characteristics of the treatment plant. Water treatment plant operators should be familiar with routine plant operations, special operations (such as startup and shutdown of individual processes), and the preventive maintenance required for each treatment process. On a daily basis, operators may be responsible for monitoring process performance, analyzing water quality (raw, settled, and finished), adjusting process controls, and inspecting plant facilities. Finally, one of the most important operator tasks is record-keeping. Maintaining a daily operation log, including keeping a diary that provides an accurate day-to-day account of plant operations, provides a historical record of events for future reference. Recording all analytical results needed to complete reports that are required by local regulatory agencies supports the regulatory compliance effort and also enables the utility to have long-term records of its water quality. Water quality monitoring for process control is discussed in chapter 3.

Membrane Filtration

Membranes are considered an alternative filtration process and consist of polymeric layers with very small pores that physically strain particles, pathogens, and so on from the influent water. Membranes are classified according to both the pore size and the amount of pressure required to force the water through the membranes. Low-pressure membranes (microfiltration and ultrafiltration) have larger pore sizes and are used for filtration, while high-pressure membranes (nanofiltration and reverse osmosis) have much smaller pore sizes and are used to modify the chemical characteristics of water being treated. Because low-pressure membranes rely on a physical removal process, i.e., straining, the size of the pores determines what contaminants can be removed from the process. Ultrafiltration membranes can remove a portion of the smaller particles that could pass microfiltration membranes. These membranes can be used as a replacement for coagulation, flocculation, sedimentation, and filtration, or can be used as a polishing step behind any combination of these processes. Because microfiltration and ultrafiltration membranes do not remove dissolved constituents such as arsenic or iron in groundwater or NOM in the form of dissolved organic carbon, some form of pretreatment such as coagulation, and perhaps clarification, may be needed prior to microfiltration or ultrafiltration if removal of dissolved substances is necessary. In this situation, information contained in this manual can be helpful for optimizing the pretreatment processes.

REFERENCES

American Public Health Association (APHA), American Water Works Association (AWWA), and Water Environment Federation (WEF). 2005. *Standard Methods for the Analysis of Water and Wastewater.* 21st ed. Washington, D.C.: APHA.

Becker, W.C., C.R. O'Melia., K.-K. Au, and J.S. Young Jr. 2004. *Using Oxidants to Enhance Filter Performance.* Denver, Colo.: Awwa Research Foundation and AWWA.

Edzwald, J.K., and J.E. Van Benschoten. 1990. Aluminum Coagulation of Natural Organic Matter. In *Chemical Water and Wastewater Treatment,* ed. H. Halan and R. Klute, 341–359. New York: Springer-Verlag.

Edzwald, J.K., and J.E. Tobiason. 1999. Enhanced Coagulation: USA Requirements and a Broader View. In *Removal of Humic Substances From Water.* Trondheim, Norway:

IAWQ/IWSA Joint Specialist Group on Particle Separation.

Logsdon, G.S., A.F. Hess, M.J. Chipps, and A.J. Rachwal. 2002. *Filter Maintenance and Operations Guidance Manual.* Denver, Colo.: Awwa Research Foundation and AWWA.

McTigue, N.E., and D. Cornwell. 1988. The Use of Particle Counting for the Evaluation of Filter Performance. In *Proc. AWWA Annual Conference, Orlando, Fla.* Denver, Colo.: AWWA.

Pernitsky, D.J., and J.K. Edzwald. 2003. Solubility of Polyaluminum Coagulants. *Journal of Water Supply: Research and Technology–AQUA,* 52(6):395–406.

Pernitsky, D.J., and J.K. Edzwald. 2006. Selection of Alum and Polyaluminum Coagulants: Principles and Applications. *Journal of Water Supply: Research and Technology–AQUA,* 55(2):121–141.

Chapter **2**

Jar Testing

Susan Teefy, James Farmerie, and Elizabeth Pyles

INTRODUCTION

The jar test is recognized throughout the water industry as a valuable tool for realistically simulating coagulation, flocculation, and sedimentation at a full-scale treatment plant. Water treatment plant operators as well as consultants, chemical suppliers, and researchers routinely perform this test. It basically involves duplicating, at bench scale, conventional treatment steps that occur in the full-scale plant. The jar test method, while duplicating the chemical treatment processes sufficiently, occasionally has limitations with regard to duplicating more advanced physical treatment processes.

Jar testing may be done for many different reasons. Tests can be performed to evaluate the effects of changes in chemical dosages and points of application; choose alternative coagulants; add polymeric coagulant aids; implement alternative preoxidation strategies; vary mixing intensities and times; and change overflow rates on the removal of particles, natural organic matter (NOM), or other water quality parameters of concern.

It is important that the conditions used in the jar test accurately simulate the full-scale plant conditions. Doing this requires knowledge of the hydraulic characteristics of the treatment steps: initial mixing (also called *rapid* or *flash mixing*), flocculation, and clarification, as well as translation into a batch-testing protocol. The key parameters include:

- Velocity gradient/mixing intensity in the rapid mix and flocculation basins

- Effective retention times in the rapid mix and flocculation basins

- Surface loading rate of the sedimentation basin

- Actual retention time in basins if jar testing is being done to evaluate time-dependent reactions for which full-scale reaction time influences results

However, even when theoretical conditions are matched closely in the jar test procedure, there is often a need to empirically tweak the parameters to make the jar test results match the full-scale results. This is commonly a result of the limitations of the jar test in matching the physical characteristics of the treatment process. Customizing a jar test procedure so it can yield results indicative of plant performance is iterative and can take a lot of time. Operators with successful jar testing procedures have often used the theoretical parameters as a starting point and then made minor adjustments by trial and error until the full-scale plant results are accurately simulated by the jar test. An example of this is presented as a case study, "Jar Test Calibration," in chapter 7.

Although jar tests are often conducted to help with full-scale plant optimization, they may also be done to meet certain regulatory requirements. Under the Stage 1 Disinfectants and Disinfection By-products Rule (D/DBPR) (1998), jar testing may be conducted as part of the "enhanced coagulation" (EC) requirements. In this case, no attempt is made to simulate the full-scale plant conditions; these EC jar tests must be done under certain predefined conditions and are used to determine alternative total organic carbon (TOC) removal requirements for a particular plant.

Velocity Gradient

The intensity of mixing is generally quantified by the velocity gradient, G, with units of s^{-1} (seconds to the minus 1 power). The velocity gradient is calculated using the energy dissipation rate in the fluid, or it can be interpolated from calibration curves. In order to attain jar test results that are relevant to the treatment plant, mixing intensities, and therefore velocity gradients, during jar tests should correspond to those in the treatment plant. Commercially available graphs provide velocity gradients in s^{-1} units relative to jar test paddle speed, water temperature, and jar volume, and examples are presented later in the text.

The velocity gradient varies significantly with water temperature (viscosity) independent of the mixing device speed. Jar tests should be conducted with water temperatures the same as or as near as possible to the plant water temperature conditions—see water baths—to simulate plant conditions, and water temperature must be recorded.

One jar test objective is to optimize velocity gradients at the rapid mix and tapered flocculation stages to achieve lower final turbidities with optimum floc development (compatible with sedimentation, direction filtration, dissolved air flotation-filtration, etc.). This should be addressed at least seasonally where significant changes in raw water temperatures occur. When the optimum velocity gradients have been determined by jar test, as depicted in Figure 2-1, the plant mixer(s) and flocculator speed/power can be adjusted according to previously prepared plant-specific information relating flocculator speed, water temperature, and G value, if such information is available. Methods for calculating G for mixing and flocculation basins are presented in chapter 4 of this manual.

Initial Mixing (Rapid Mix)

When a jar test is performed, the rapid mixing time should be equivalent to the retention time of the mixing chamber in the treatment plant, and G values should match the full-scale conditions, if possible. Typical mixing intensities for a well-designed flash mix basin range from 700 to 1,000 s^{-1}, and in a typical jar test, times of 30–60 sec are used, with high paddle speeds of 100–300 rpm. If rapid mixing in the plant is done by in-line mixers or by chemical induction mixers, simulating the full-scale mixing conditions may not be possible using typical 2-L jars and jar test equipment.

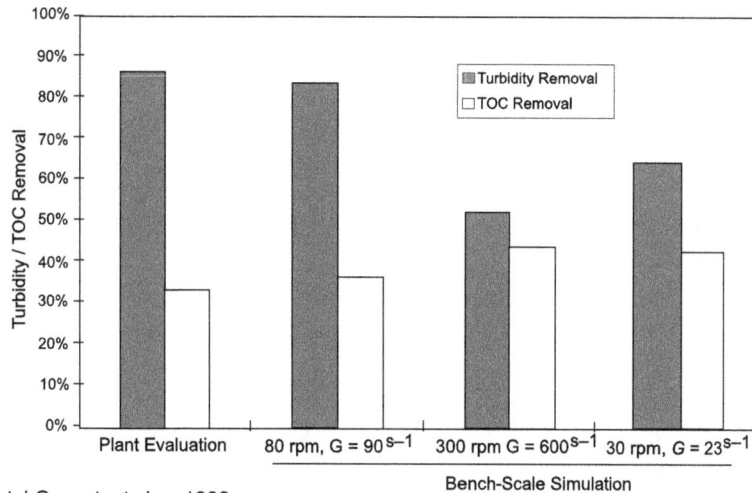

Source: AH Environmental Consutants Inc. 1998.

Figure 2-1   Use of jar tests to determine optimum flash mix conditions

G-value testing requires repeated jar tests. Preferably chemical coagulation conditions for alkalinity, pH, and coagulant dosage should be determined first, followed by testing different G values in a series of jars, all of which have received the same chemical treatment and have water at the same temperature as the influent water in the plant.

Flocculation

In jar testing, the retention time and G values of the flocculation basin should correspond to those in the water treatment plant. It may be necessary to vary the mixing speeds, if the plant performs tapered flocculation in different compartments. Typical retention times are approximately 15–30 min, while typical mixing intensities vary from 10 to 40 s^{-1}, with the more intense mixing steps occurring first, followed by the gentler mixing. Users of jar tests for evaluating flocculation need to consider that in jar tests, flocculation is done on a batch basis for a specific retention time, whereas retention times for floc particles in continuous-flow plant-scale flocculation basins vary over a range because of short-circuiting that inevitably occurs in those basins.

Sedimentation

A key jar test objective for evaluating sedimentation processes is to optimize floc development and settleability. The settling velocity of floc must be higher than the surface loading rate of the clarification basin, or the floc will not settle. The surface loading or overflow rate is determined by dividing the basin flow rate by the surface area, commonly expressed in either gallons per day per square foot (gpd/ft^2) or gallons per minute per square foot (gpm/ft^2). A basic unit conversion reveals that the surface loading corresponds to upflow velocity:

$$\text{Surface loading rate} = \frac{\text{Flow rate}}{\text{Surface area}} = \frac{\text{Volume} \div \text{Time}}{\text{Area}} = \frac{\text{Length}}{\text{Time}} \qquad \text{(Eq 2-1)}$$

or

$$\frac{\text{gpd}}{\text{ft}^2} \times 3{,}785\,\frac{\text{cm}^3}{\text{gal}} \times \frac{1\,\text{day}}{1{,}440\,\text{min}} \times \frac{1\,\text{ft}^2}{929\,\text{cm}^2} = \frac{\text{cm}}{\text{min}} \qquad \text{(Eq 2-2)}$$

Table 2-1   Settling velocity conversion factors for clarification basins

Units for Plant Flow Rate/Surface Area	Multiply by	Settling Velocity
mgd/ft^2	2,830	cm/min
gpm/ft^2	4.07	cm/min
gpm/ft^2	2.44	m/hr
gpm/ft^2	1.6	in./min

When sedimentation is used for clarification, an important characteristic of floc is its settling velocity. Typical surface loading rates for conventional sedimentation basins are approximately 0.5 to 1.0 gpm/ft^2 (1.2 to 2.4 m/hr) or about 2 to 4 cm/min. This can be compared to a "good" jar test sedimentation rate of >2.5 cm/min for metal hydroxide floc formed without polymer.

To calculate the settling velocity from the plant flow rate and the total sedimentation basin area, the conversion factors in Table 2-1 can be used.

Theoretically, a sedimentation basin will remove all particles that exceed this critical velocity for a given overflow rate. This concept is most important for simulating the sedimentation process in a jar test. The retention time of the full-scale settling basin must not be used in the jar test to evaluate settleability of floc. In order to obtain useful results, the jar's surface loading rate, which corresponds to a settling velocity, should closely match that of the process. In jar testing, this match is accomplished by collecting a sample from the jar at a set depth below the water surface at a given time.

For example, a typical sedimentation basin is designed for an overflow rate of 0.5 gpm/ft^2. This value corresponds to a particle settling velocity of 2 cm/min. If the sampling port of the jar is 10 cm below the water surface, then all particles that pass the sampling port within 5 min (10 cm divided by 2 cm/min) would be removed. Therefore, samples should be collected 5 min after the flocculation period to simulate the performance of the sedimentation basin. If the jar were sampled on the basis of the settling basin's retention time (e.g., 120 min), the performance of the basin would be grossly overestimated.

Figure 2-2 illustrates a minimal effect of polymer addition for samples taken after 10 min. When the same plant is operating at a maximum surface loading rate of 1,800 gpd/ft^2, which corresponds to a sampling time of 2 min, the effect is more pronounced.

Standard jars used in jar tests have sample ports that are 10 cm below the water surface when using the full capacity of the jar (i.e., 2 L in a 2-L jar). For these jars, the sampling time is determined by dividing the 10 cm by the settling velocity, in cm/min. Typical sample collection times in the jar test are between 2 and 5 min.

Points of Chemical Applications

If the purpose of a jar test is to optimize conditions for an existing full-scale plant, chemicals must be added in the same order as in the plant. Alternatively, the same chemicals can be added at different times or in a different order in an attempt to improve the quality of water produced. For example, a jar test might evaluate the potential effect of polymer addition ahead of the primary coagulant or the effect of adding caustic for pH adjustment before adding alum versus after adding alum.

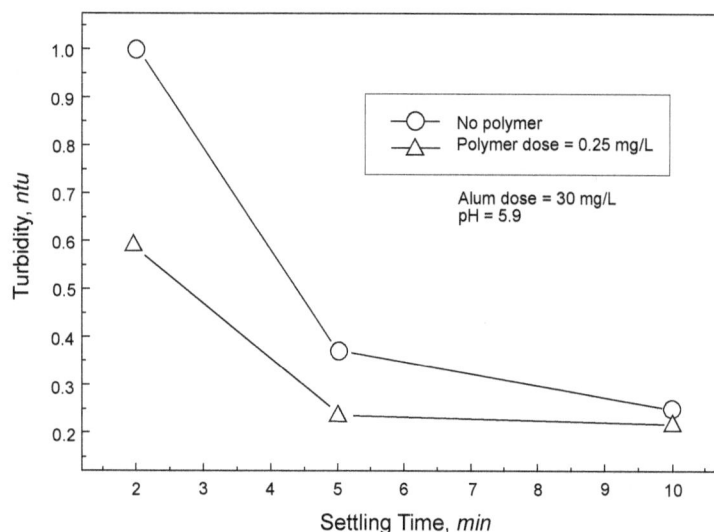

Source: AH Environmental Consultants Inc. 1997.

Figure 2-2   Example use of jar tests: Settled turbidity versus settling time

PREPARING FOR A JAR TEST

The development of a realistic and useful jar test protocol depends on several major steps. These steps require time and effort on the part of the operator, but the benefit will be well worth the effort. A well-planned and organized approach will save time in the long run by reducing guesswork and identifying the critical resources and information needed to ensure that a jar test yields useful data.

Defining Study Goals

Jar tests can be conducted for a number of reasons, so it is critical to define the objective of the test before beginning. The most common use of the jar test is for day-to-day process control, but it may also be used for other reasons. For example, goals might include

- Determining the point of diminishing returns (PODR) or the optimum coagulation pH for TOC removal under the Stage 1 D/DBPR

- Evaluating alternative chemicals such as coagulants (ferric chloride, aluminum sulfate, polyaluminum chloride, etc.), pH controls (lime versus caustic soda), or preoxidants (potassium permanganate, ozone, chlorine dioxide, etc.)

- Assessing additional chemical choices; for example, organic polymers for enhanced solids-liquid separation or effect of powdered activated carbon for taste and odor removal

- Evaluating physical modifications, such as varying mixing intensities and points of chemical application, implementing tapered flocculation, or installing baffles to increase retention times or possible procedures for new physical facilities

Water quality parameters that are commonly used to assess treatment performance include

- Turbidity and color removal

- Dissolved organic carbon (DOC) concentration or ultraviolet absorbance at 254 nm

- Disinfection by-product formation

- Iron and manganese removal

- Taste and odor of the treated water

- Sludge characteristics

- Algae concentrations

- Residual aluminum levels

- pH adjustment

Required Information

The next step in preparing a useful jar test protocol is to gather the necessary information about the physical characteristics of the plant, chemical application points, and current full-scale plant performance. To make valid comparisons between jar test data and actual plant performance data, an operator needs to know the key parameters of the existing or future processes. These factors should include velocity gradients or mixing intensities and detention times for mixing, flocculation, and channels or pipes, as well as overflow rates for sedimentation basins.

Velocity gradient. The design documents or operations and maintenance manual for a water treatment plant should provide data for actual in-plant velocity gradients for the rapid mix and the flocculation basins. Figure 2-3 shows an example of a chart relating velocity gradient to flocculator speed and water temperature. The accuracy of this information is very important for reliable results. If the information is not readily available, use the procedure for calculating velocity gradient described in chapter 4.

Detention times. The theoretical detention or retention time is defined by the following equation:

$$\text{Detention time} = \frac{\text{Tank volume}}{\text{Flow rate}} \qquad \text{(Eq 2-3)}$$

For example, the detention time of a tank with the dimensions 7 ft (2.14 m) × 6 ft (1.83 m) × 10 ft (3.06 m) (depth × width × length) operated at a flow rate of 105 gpm (23.9 m³/h) would be

$$\frac{7\,\text{ft} \times 6\,\text{ft} \times 10\,\text{ft}}{105\,\dfrac{\text{gal}}{\text{min}}} \times 7.48\,\frac{\text{gal}}{\text{ft}^3} = 30\,\text{min} \qquad \text{(Eq 2-4)}$$

$$\frac{2.14\,\text{m} \times 1.83\,\text{m} \times 3.06\,\text{m}}{23.9\,\dfrac{\text{m}^3}{\text{hr}}} \times 60\,\frac{\text{min}}{\text{hr}} = 30\,\text{min} \qquad \text{(Eq 2-5)}$$

In this way, theoretical detention times can be readily calculated from fixed basin dimensions and the flow rates through each basin. However, effective detention times through a plant's basins may be less than half of the theoretical detention

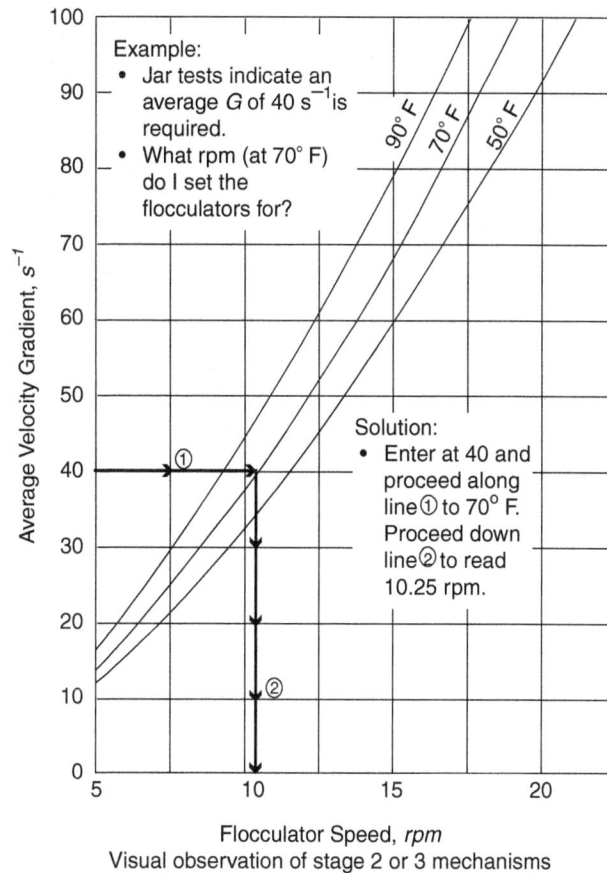

Courtesy of J. Edward Singley and Tim Brodeur.

Figure 2-3   Example use of graph for determining velocity gradient for jar test based on full-scale

times, perhaps because of short-circuiting in the process units. Special studies may need to be done to characterize the effective detention time; see chapter 4 for trouble-shooting tips on short-circuiting. Finally, make sure to take into account not only the size of each basin but also the number of basins in service when dividing the volume by the flow rate.

Sedimentation basin overflow rate.   The sedimentation basin overflow rate, or surface loading, determines the length of time that the water is allowed to settle in a jar before a sample is taken. Standard practices for jar testing include collection of turbidity samples at various times after stirring has stopped in order to simulate full-scale plant turbidity removal performance. However, the contact time in the full-scale basin may be 2 to 4 hr. The result is that for a given settling velocity, the time of sedimentation in the jar test is about one-hundredth that in the conventional plant. Such short holding times, however, are not adequate for assessing chemical reactions that may be occurring in the basin, such as formation of disinfection by-products or for evaluating the efficacy of powdered activated carbon (PAC). If the purpose of the bench testing procedure is to determine a reaction time-dependent parameter during the treatment process, it will be necessary to hold the jars for the actual retention time in full-scale process basins rather than to take samples at shorter times related to the basin overflow rate.

Chemicals and points of application.   Another planning step involves defining which chemicals to use in the jar test. If the objective of the test is to assist in

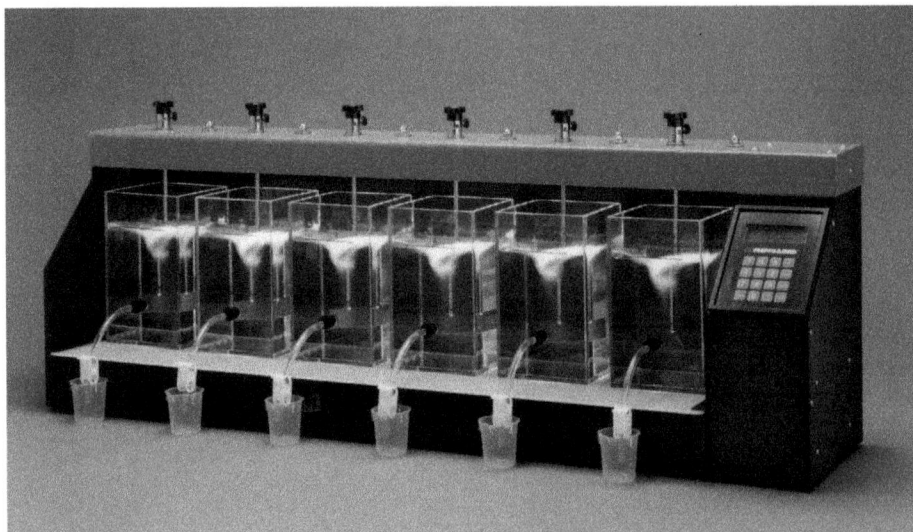

Source: Phipps & Bird, Inc. Richmond, Va.

Figure 2-4   Example jar test unit

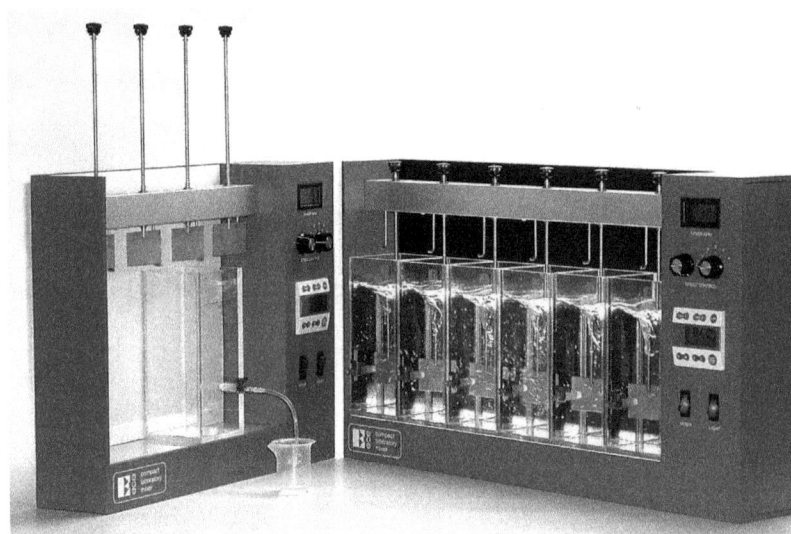

Source: EC Engineering Inc.

Figure 2-5   Example jar test units, 4-jar and 6-jar systems

day-to-day plant optimization, then the chemicals that are currently used in the treatment process should be used in the jars, and in the same order that they are used in the plant. If alternative chemicals are to be evaluated, these must be used in the jars. The operator should have data sheets for each chemical with information about the chemical formula, specific gravity, percent weight, viscosity for liquids, solubility for solids, and safety information. If the chemical being evaluated in a test has never been used in the process, the previous information should be supplemented with data for unit cost, shipping, and storage requirements, as well as a current material safety data sheet (MSDS).

Current treatment performance data.   In order to determine how well the jar testing conditions or procedures simulate conditions in the full-scale facility, the operator should obtain treatment plant performance data. This information will be

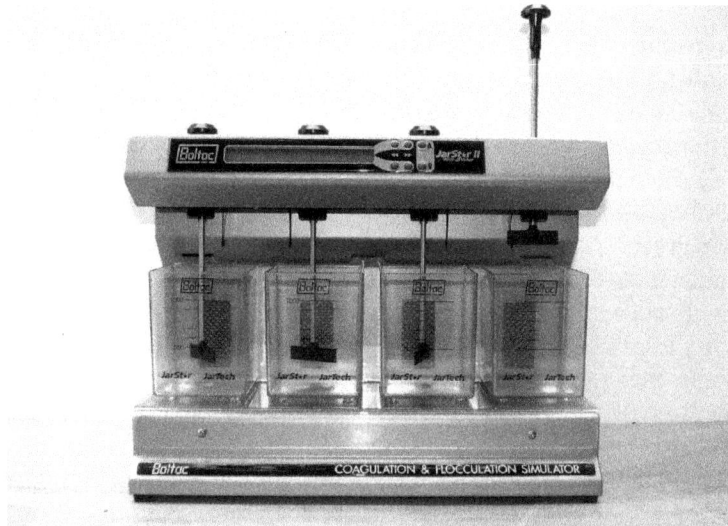

Source: Boltac Industries, New Zealand.

Figure 2-6   Example jar test unit

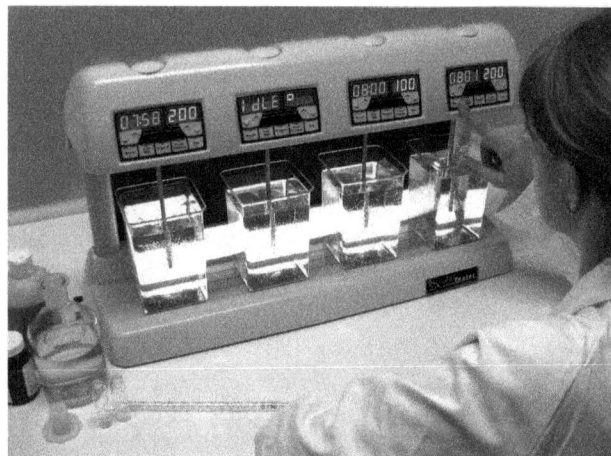

Source: Aquagenics Pty Ltd.

Figure 2-7   Example jar test unit

compared to that from the jar tests conducted under the same conditions. Water quality data from the plant's settled water are generally used, such as the settled water turbidity, pH, and TOC or ultraviolet absorbance values.

Equipment

Jar testers.   Jar test equipment consists of jars to hold the water, the impeller, a mechanism to turn the impeller, and lab equipment to analyze the results. The following sections describe each of these items and the role of each in the jar test procedure. Examples of typical jar test apparatus are shown in Figures 2-4 through 2-7. Typically, these devices includes four or six jars with sample ports and paddle mixers, and can be preprogrammed to stir at particular speeds for particular amounts of time, simulating the coagulation, flocculation, and sedimentation processes.

Types of jars.   Various types of jars have been used, including cylindrical 1-L and 2-L glass beakers. The most commonly used jars today are 2-L square beakers.

The square plastic beakers provide better mixing conditions than round beakers; therefore the remainder of this section will focus on these containers. Commercially available square jars are fabricated by cementing together clear acrylic sheets or by injection molding. These jars are often referred to as gator jars (Cornwell and Bishop 1983) because they were developed at the University of Florida. Desirable attributes for the jars include transparency, volume markings, appropriate sampling point, ease of cleaning, and durability.

Stirrer. A stirring mechanism turns the impellers. Two basic types of stirring mechanisms are used: gear-driven and magnetically driven units. A gear-driven unit has a variable-speed motor that turns four to six gears and stirrer shafts. The motor(s) and gears are located above the jar test containers so that the shafts for the impellers extend down into each jar. Units are available that can be controlled from about 10 to 300 rpm. This is the type utilized in the standard jar test apparatus shown in Figure 2-4 through Figure 2-7.

The magnetically driven unit works on the same principle as a magnetic stirrer plate. The paddle contains a magnet that turns as the metal under it rotates. Either type of unit may be used for jar testing. The magnetically driven unit has the advantage of providing open space above the jars to add chemicals. It is also possible to construct a jar stirrer using one or more variable-speed mixers. Units can be purchased that operate over a broad rpm range, allowing for an intensive rapid mix. They offer the advantage of testing different mix intensities simultaneously because the stirrers are individually controlled. It may also be less expensive to buy three or four mixers and mount them on ring stands than to purchase commercially available units.

Impellers. Several types of impellers are available: paddle type, turbine, marine, and axial flow. Ideally, the impeller that best simulates the full-scale process should be used. Interchanging impellers requires a stirrer mechanism equipped to allow replacement of the shaft/impeller.

Magnetic stirrers are available only with flat-paddle impellers and would be difficult to retrofit with a different impeller type. They can, therefore, effectively simulate the action of a paddle, walking-beam, or flat-blade turbine types of mixing devices. Figure 2-8 shows the velocity gradients for a standard magnetic stirrer in a 2-L square jar.

Standard top-mounting stirrers are generally provided with flat paddles. However, the shaft and impeller can be changed to allow testing of different impeller types. Figures 2-8 through 2-10 show G values for various combinations of jars and impellers.

Water bath. A water bath is a tank in which the jars sit during the jar test as water circulates around them to maintain the correct water temperature. The use of a water bath is not usually required unless the treated water is very cold and the jar test is to be done in the warm laboratory. A simple test determines if a water bath is necessary: Take a sample of the raw water and immediately run a jar test; take another sample of the same water and let it warm up a few degrees, then conduct the same test and note if the results differ. Cold waters are frequently difficult to treat because lower temperatures result in water that is lower in viscosity, higher in density, and with slower chemical reaction times for flocs to form. Therefore, even a slight temperature increase could make treatment easier than in the full plant, causing misleading results.

Analytical equipment and laboratory ware. A variety of lab equipment is required to prepare standardized solutions and analyze water samples. The most essential analytical equipment includes a turbidimeter, pH meter, and thermometer. Regulatory requirements may require operators to perform more advanced analytical

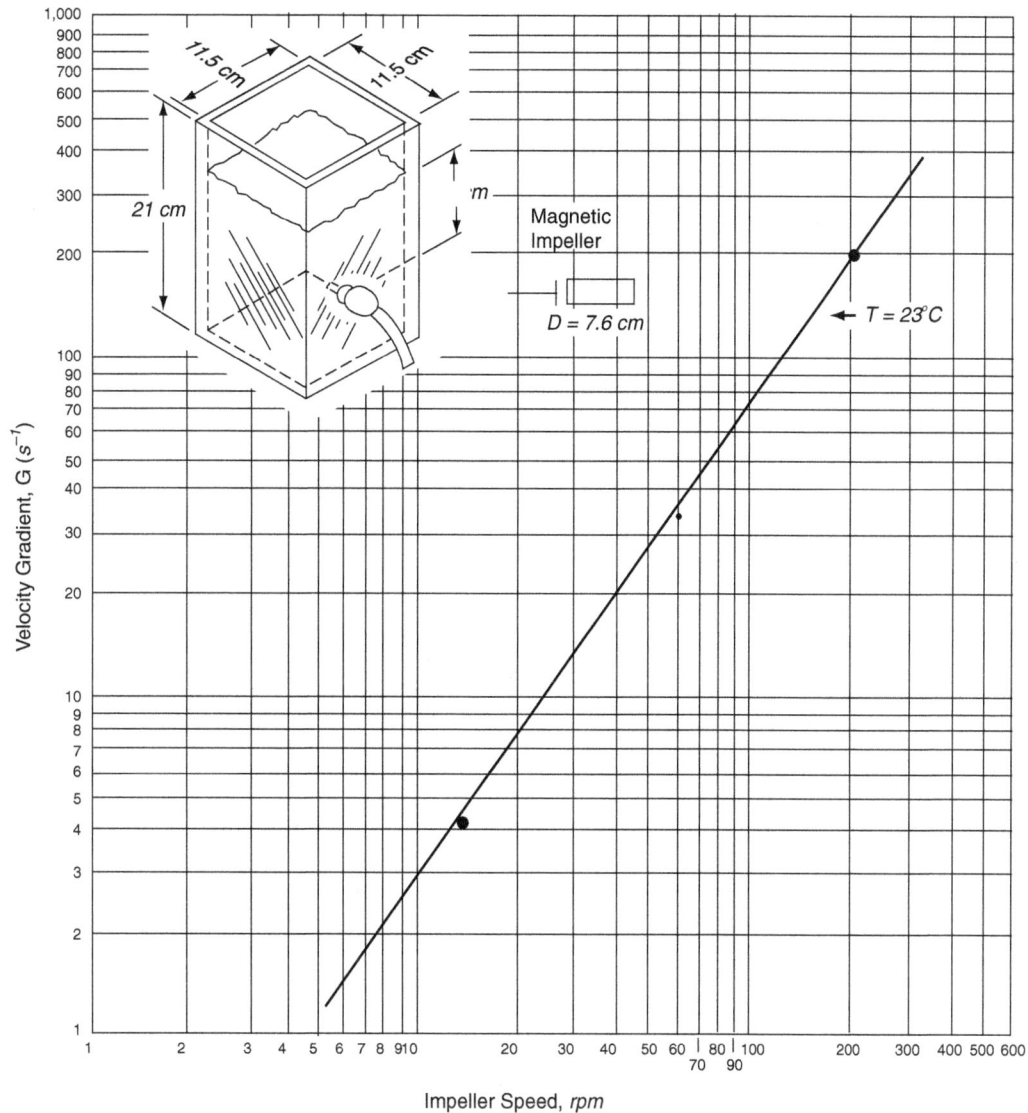

Source: Environmental Engineering & Technology Inc.

Figure 2-8   Laboratory *G* curve for magnetic jar tester with gator jar

analyses, which may call for the use of a total organic carbon analyzer or ultraviolet (UV) spectrophotometer. If the analytical equipment is not available, the operator must make arrangements with a commercial laboratory for needed tests. The laboratory may also provide suitable sample containers. Zeta potential meters are sometimes used to assess results of the jar test, as are particle counters.

An assortment of pipettes, burettes, syringes, graduated cylinders, volumetric flasks, and a laboratory weight scale may also be required to make the necessary solutions. If chemicals are to be added without dilution, a micropipette is required. Many operators use PTFE septa (small, flat disks originally intended as liners of glass sample bottles) for holding minute volumes of neat chemical until ready for use. When needed, they are simply tossed into the jar along with the neat chemical, where they quickly sink to the bottom as the chemical is being mixed into the water.

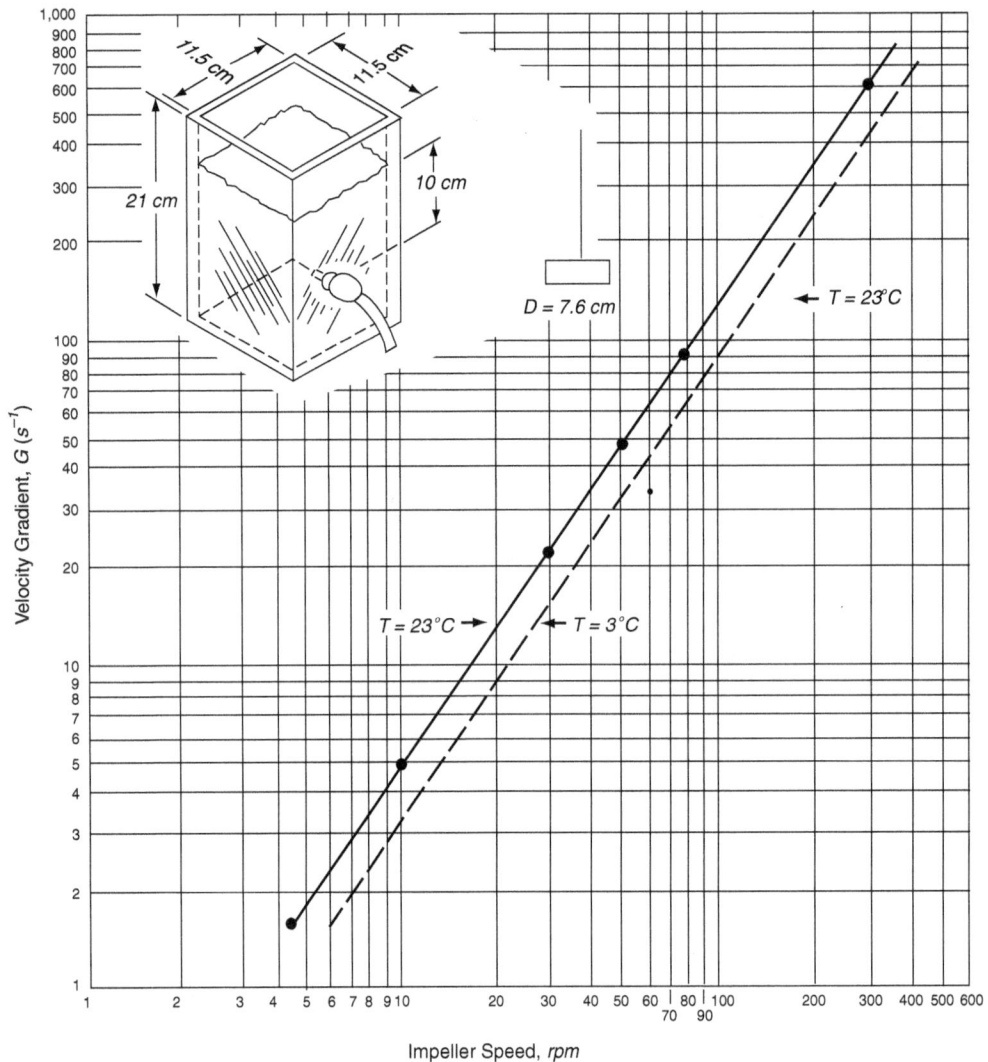

Source: Cornwell and Bishop 1983.

Figure 2-9   Laboratory G curve for flat paddle in the gator jar

Tools for simultaneous chemical addition.   When comparing various chemical dosages in a jar test, it is necessary to add the chemicals to all jars simultaneously. Under most conditions, manually adding the chemical quickly to each jar is acceptable. However, when small time changes are important, it is necessary to be able to dose all jars at exactly the same time. For example, if a test involves a short mix time of just a few minutes, and it takes 1 min to dose all six jars, the first jar dosed will have a significantly longer reaction time than the last jar. In such cases, operators have built devices such as the cup holder (Figure 2-11) and the septa bar (Figure 2-12) for holding small cups filled with the premeasured chemicals. Such devices allow dosing of all jars simultaneously.

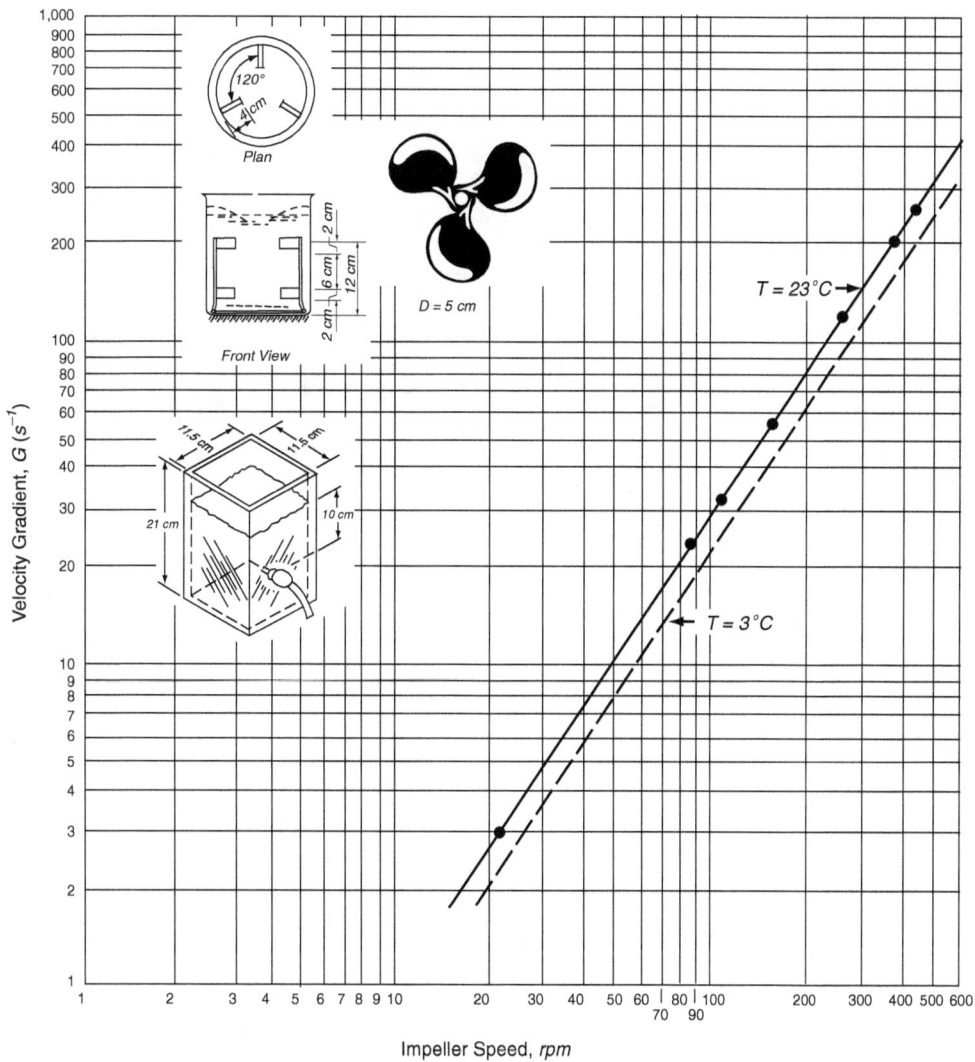

Source: Cornwell and Bishop 1983.

Figure 2-10 Laboratory *G* curve for marine propeller in either the Hudson or gator jar

Chemicals

The chemicals used for pretreatment processes generally fall into one of four categories:

- Coagulants
- Coagulant aids
- pH controllers
- Oxidants/disinfectants
- Adsorbents (e.g., activated carbon)

Source: Zone 7 Water Agency.

Figure 2-11   Wooden holder for six dosing cups: Small plastic cups fit into each hole drilled in the wood

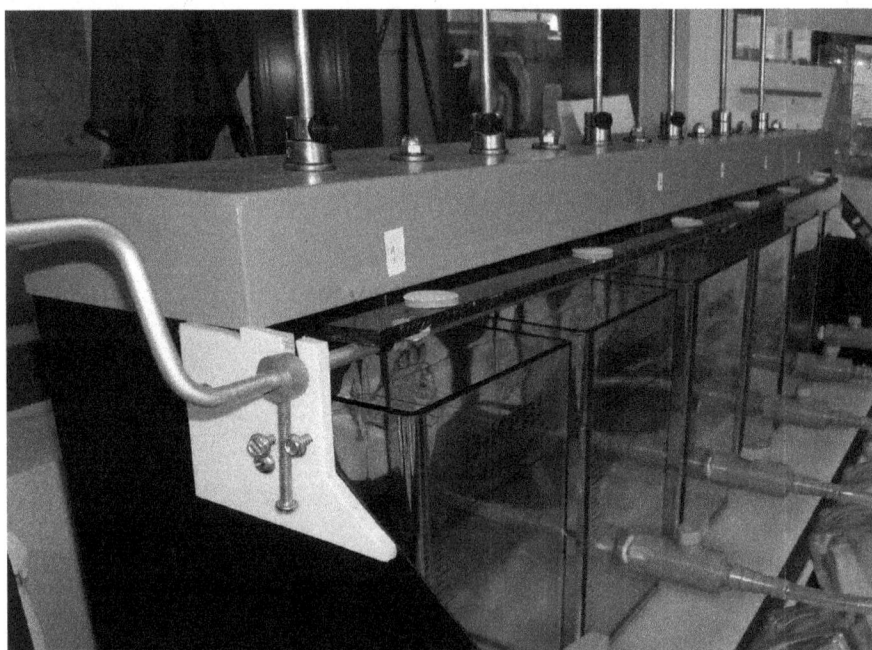

Source: Alameda County Water District.

Figure 2-12   Septa bar: metal bar to hold multiple septa over jars until ready to dump septa into all jars simultaneously.

In addition to those listed, other chemicals may be needed during and after the jar test. Some of the chemicals listed are available as pure substances of uniform quality. Depending on their origins, others may vary in chemical composition because of their complexity, causing differences in effectiveness. Polymers are an example. In order to achieve useful results, jar testing must use treatment chemicals that are being used or will be used at the full-scale plant and are certified for use in drinking water treatment (NSF/ANSI 60 [NSF International and American National Standards Institute 2005]).

Coagulants. Coagulants and how they work are discussed in detail in chapter 1. The most commonly used coagulants include

- Aluminum sulfate (alum)

- Ferric salts (ferric chloride and ferric sulfate)

- Ferrous sulfate

- Polymeric inorganic coagulants (partially neutralized metal salts such as polyaluminum chloride)

Expressing coagulant concentrations. When a jar test is performed, the correct dosages of chemicals must be applied. The operator must understand whether the applied dosage is stated "as product" or "as ingredient" basis. One way to avoid confusion is to express dosages/concentrations in terms of the molar amount of active ingredient, i.e., the moles of iron or aluminum per liter of solution. This convention is of even greater importance for jar tests comparing different coagulants. Equimolar dosages of aluminum and iron contain the same number of atoms. To determine the molar metal dosage applied at the full-scale treatment plant, data for the plant flow rate, bulk chemical feed rate, and the chemical data sheet are needed. The data sheets for most liquid aluminum or iron-based coagulants report values for the specific gravity and percent Al_2O_3 or percent Fe. To convert to molar metal dosage, apply the formula:

$$\text{Dosage}, \frac{\mu mol}{L} = \frac{\text{Chemical feed rate}, \dfrac{lb}{day}}{\text{Plant flow rate}, mgd \times 8.34} \times \text{Factor} \qquad \text{(Eq 2-6)}$$

The "factor" depends on the coagulant used, as shown in Table 2-2. The molar metal dose at the full-scale plant obtained in this way can then be used as a baseline value for the jar test.

Table 2-2 Factors for determining molar metal dosage

Coagulant	Factor
Dry alum ($Al_2(SO_4)_3 \cdot 14H_2O$)	3.365
Liquid aluminum product (alum, PACl, etc.)	%Al_2O_3 by weight ÷ 5.1
Dry ferric chloride (anhydrous, $FeCl_3$)	6.165
Dry ferric sulfate (anhydrous, $Fe_2(SO_4)_3$)	5.004
Dry ferrous sulfate (anhydrous, $FeSO_4$)	6.579
Liquid iron product (ferric sulfate, polyferric chloride, etc.)	%Fe by weight ÷ 5.58

Example: A treatment plant feeds 5,000 lb (2,270 kg) of liquid alum per day at a flow rate of 5 mgd (19 ML/d). According to the data sheet, the product contains 8.3 percent Al_2O_3 by weight. Therefore the molar-equivalent dosage equals:

$$\frac{5,000}{(5\times8.34)}\times\frac{8.3}{5.1}=195\ \mu M\left(\frac{\mu mol}{L}\right)\ \text{as Al} \tag{Eq 2-7}$$

If the operator were to test ferric chloride as an alternative coagulant, a comparable dosage to the currently applied alum would be 195 µM as Fe. Note that 195 µM as Al corresponds to 58 mg/L as alum; ferric chloride at a dosage of 58 mg/L as Fe would be five times the molar Al concentration. Figure 2-13 illustrates the relationship of molar metal concentration to alum or iron concentration in mg/L.

Making up stock solutions from the dry chemical. During jar testing, it is often inconvenient to feed a dry chemical to the jars. To test dry metal salts, stock solutions should be made up according to the following instructions:

Dry alum: Dissolve 10.0 g of dry aluminum sulfate ($Al_2(SO_4)_3 \cdot 14H_2O$) in distilled or deionized water and dilute to the 1,000-mL mark in a volumetric flask. The

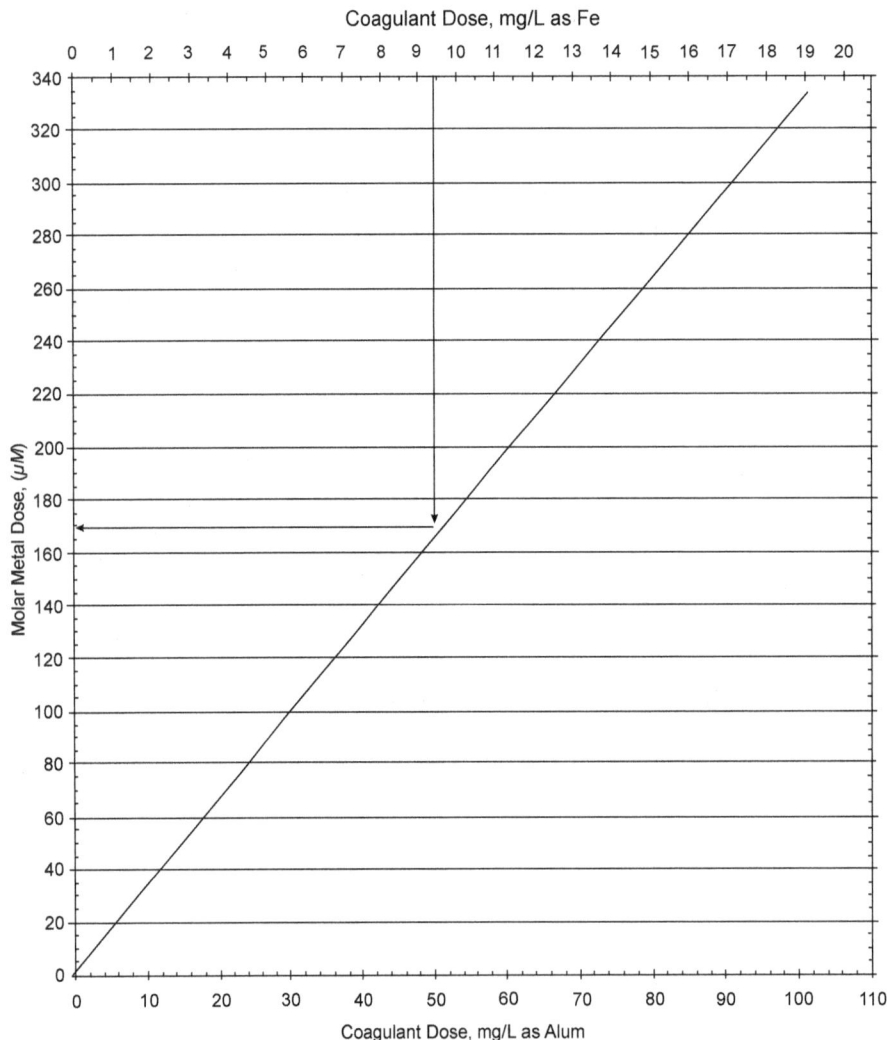

Figure 2-13 Expressing coagulant doses in molar metal concentrations

resulting solution contains 10,000 mg/L alum, which corresponds to 0.17 percent Al_2O_3 by weight. Adding 1.0 mL of this stock solution to a 2-L jar results in a dose of 5 mg/L as alum or 17 μM as Al.

Dry ferric chloride (anhydrous).   Dissolve 2.93 g of dry ferric chloride (anhydrous, $FeCl_3$) in distilled or deionized water and dilute to the 1,000-mL mark in a volumetric flask. The resulting solution has a concentration of 1,000 mg/L or 0.10 percent Fe by weight. Therefore, adding 1.0 mL of stock solution to a 2-L jar results in 0.50 mg/L or 9.0 μM as Fe.

Dry ferric sulfate (anhydrous).   Dissolve 3.57 g of dry ferric sulfate (anhydrous) in distilled or deionized water and dilute to the 1,000-mL mark in a volumetric flask. The resulting solution has a concentration of 1,000 mg/L or 0.10 percent Fe by weight. Adding 1.0 mL of stock solution to a 2-L jar results in 0.50 mg/L or 9.0 μM as Fe.

Dry ferrous sulfate (anhydrous).   Ferrous sulfate works in much the same way as ferric sulfate, except that it provides a bivalent iron ion (Fe^{2+}) when dissolved in water. Generally, the ferrous iron is oxidized to the trivalent form (Fe^{3+}) through chlorination prior to use. To oxidize ferrous sulfate to ferric iron, add 0.13 mg of chlorine for each milligram of ferrous sulfate. To prepare a stock solution, dissolve 2.7 g of ferrous sulfate (anhydrous) in distilled water and dilute to 1,000 mL. The resulting solution has a concentration of 1,000 mg/L or 0.10 percent Fe by weight. Adding 1 mL of stock solution to a 2-L jar results in 0.50 mg/L or 9.0 μM as Fe.

Polymeric inorganic coagulants.   Note that the effectiveness of dry and liquid coagulants varies with dilution, and highly diluted stock solutions may degrade over time. Thus, new stock solutions must be prepared every day. Polymeric inorganic coagulants such as polyaluminum chloride can degrade very quickly and should not be diluted but used neat.

Working with liquid coagulants.   Dilution of liquid coagulants is usually not required if the necessary volume can be added to test jars with a micropipette, an

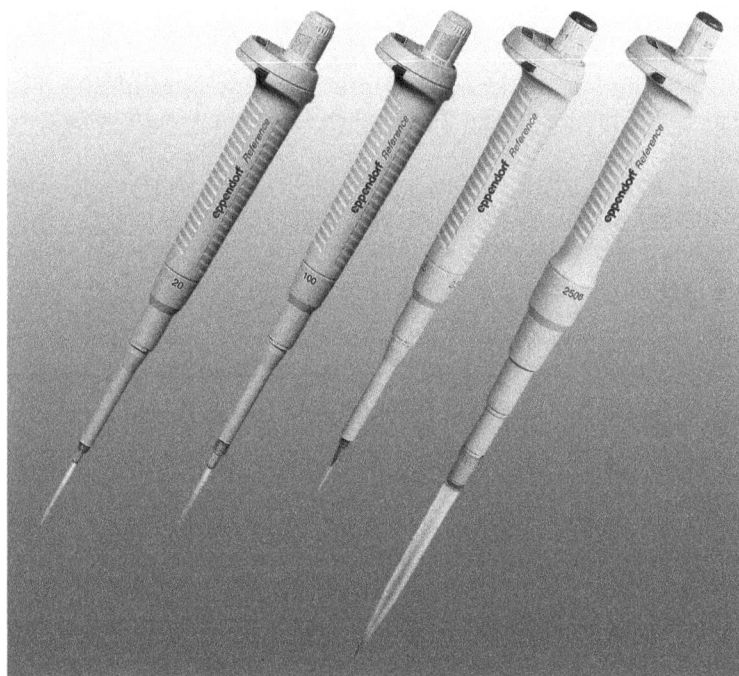

Source: Eppendorf North America.

Figure 2-14   Example of micropipettes capable of dispensing 0.1 μL to 2,500 μL

example of which is shown in Figure 2-14. If a micropipette is not available, dilution at a ratio of at least 1:100 by volume with deionized or distilled water is recommended.

The amount in microliters (µL) to add to a 2-L jar can be calculated from the percent Fe or percent Al_2O_3 content and the specific gravity of the coagulant as follows:

For aluminum products:

$$\text{Volume, µL} = \frac{\text{Dose, µM} \times 5.10 \times \text{Volume of jar, L}}{\%Al_2O_3 \times \text{Specific gravity}} \times \text{Dilution factor} \quad \text{(Eq 2-8)}$$

For iron products:

$$\text{Volume, µL} = \frac{\text{Dose, µM} \times 5.59 \times \text{Volume of jar, L}}{\%Fe \times \text{Specific gravity}} \times \text{Dilution factor} \quad \text{(Eq 2-9)}$$

Example: A water utility tests two alternative coagulants, a polymeric ferric coagulant (sp gr 1.49, 12.27 percent Fe) and aluminum chlorohydrate (sp gr 1.34, 23.5 percent Al_2O_3). The plant currently uses anhydrous ferric chloride at a dosage of 9.4 mg/L as Fe. Using Eq 2-8, the current dosage corresponds to a molar metal concentration of 168 µM. To evaluate the effectiveness of these chemicals at equimolar concentrations, the following amounts have to be added to a 2-L jar:

Aluminum chlorohydrate:

$$\frac{168 \times 5.1 \times 2}{23.5 \times 1.34} \times 1 = 54 \text{ µL} \quad \text{(Eq 2-10)}$$

Polyferric:

$$\frac{168 \times 5.59 \times 2}{12.27 \times 1.49} \times 1 = 103 \text{ µL} \quad \text{(Eq 2-11)}$$

Chemical dosages are commonly expressed as mg/L "as product" by many operators, rather than as molar equivalents. When dosages are used in terms of mg/L, the number of µL to add to the 2-L beaker for each 1 mg/L dosage can be calculated as follows:

$$\text{Number of µL to add per } \frac{\text{mg}}{\text{L}} \text{ of product} = \frac{2}{\text{Specific gravity} \times \% \text{ Active ingredient}}$$

Example: An operator wishes to dose a jar with 17 mg/L of ferric chloride. This product has a specific gravity of 1.47, and 43 percent active ingredient. The following calculation applies:

$$\frac{2}{1.47 \times 0.43} = 3.16 \text{ µL of } FeCl_3 \text{ per } \frac{\text{mg}}{\text{L}} \text{ dose}$$

So for a dose of 17 mg/L,

$$3.16 \text{ µL} \times 17 = 53.7 \text{ µL}$$

This amount can be dispensed directly onto the septa using a micropipette.

If a micropipette is not available to dispense these small volumes, add 10 mL of the coagulant to a 1-L volumetric flask and dilute to the 1-L mark with distilled or deionized water (dilution factor = 100). The volumes to be added to a 2-L jar would then be:

Aluminum chlorohydrate:

$$\frac{168 \times 5.1 \times 2}{23.5 \times 1.34} \times 100 = 5,400 \ \mu L = 5.4 \ mL \qquad \text{(Eq 2-12)}$$

Polyferric:

$$\frac{168 \times 5.59 \times 2}{12.27 \times 1.49} \times 100 = 10,300 \ \mu L = 10.3 \ mL \qquad \text{(Eq 2-13)}$$

As an alternative, when using standard mg/L concentrations of coagulants (not molar equivalents), stock solutions are often made in such a way that uniform increments can be added to the jars to achieve the desired dosages. For example, to prepare a solution that results in a 5-mg/L dosage for each 1 mL added to the 2-L jar, the following calculations apply:

$$\text{Stock solution concentration, \%} = \frac{\text{Desired dosage, } \frac{mg}{L} \times \text{Volume of jar, mL}}{\text{mL of Stock solution added to each jar} \times 10,000 \ (\frac{mg}{L} \text{ to \% conversion})} \qquad \text{(Eq 2-14)}$$

For this example,

$$\frac{5 \frac{mg}{L} \times 2,000 \ mL}{1 \ mL \times 10,000} = 1.0 \ \% \text{ Stock solution} \qquad \text{(Eq 2-15)}$$

Therefore, if a 1.0 percent stock solution is made, every 1.0 mg/L of this solution added to the 2-L jar results in a dosage of 5.0 mg/L of the product being tested.

The product strength, which is the mass of the active ingredient per gallon, also must be known. This value, in lb/gal, is often provided by the chemical supplier for each load of material delivered. If it was not provided directly, it can be calculated from the specific gravity of the material and the percent active ingredient, both of which are generally provided by the manufacturer. Specific gravity can be verified by using a hygrometer, or it can be measured directly using a graduated cylinder and a scale. Some manufacturers will provide tables relating the specific gravity to the percentage of active ingredient for a particular coagulant. The following equation is for the product strength, in lb/gal:

$$\text{Strength, } \frac{lb}{gal} = \% \text{ Active ingredient} \times \text{Specific gravity} \times 8.34 \frac{lb}{gal} \qquad \text{(Eq 2-16)}$$

When the product strength and the desired percent stock solution to be prepared are determined, the following equation can be used to determine the number of mL of neat product to add to the deionized water to make the working solution:

$$\text{Volume of coagulant, mL} = \frac{\% \text{ Stock solution} \times \text{Flask volume, mL} \times 8.34 \frac{lb}{gal}}{100 \times \text{Product strength, } \frac{lb}{gal}} \qquad \text{(Eq 2-17)}$$

In the example above, using alum with a specific gravity of 1.22 and 48 percent active ingredient, to make 500 mL of solution for the jar tests, the calculations are:

$$\text{Strength, } \frac{lb}{gal} = 0.48 \times 1.22 \times 8.34 \frac{lb}{gal} = 4.88 \frac{lb}{gal} \qquad \text{(Eq 2-18)}$$

$$\text{Volume of alum, mL} = \frac{1.0 \times 500\,\text{mL} \times 8.34\,\dfrac{\text{lb}}{\text{gal}}}{100 \times 4.88\,\dfrac{\text{lb}}{\text{gal}}} = 8.5\,\text{mL} \qquad \text{(Eq 2-19)}$$

Therefore, add 8.5 mL of neat liquid alum to the 500-mL volumetric flask and fill to the mark with deionized or distilled water. This solution would be very convenient for testing alum dosages of 5 mg/L, 10 mg/L, 15 mg/L, etc. in the 2-L jars.

Coagulant aids. In a most general sense, a coagulant aid is any substance used in conjunction with a primary coagulant, such as alum, to assist coagulation. By far the most significant coagulant aids are the synthetic organic polymers. Polymers are chains of small subunits or monomers, and they may contain ionizable groups. Depending on the charges of the functional groups on the monomeric units and the molecular weights, they are classified as cationic, anionic, or nonionic polymers and of low, medium, or high molecular weight.

Guidance for the evaluation and selection of polymers is provided in the Awwa Research Foundation (now Water Research Foundation) publication *Procedures Manual for Selection of Coagulant, Filtration, and Sludge Conditioning Aids in Water Treatment* (Dentel 1986).

Preparation and use of polymers. Polymers are available as liquid, emulsion, and dry powder products. The following paragraphs include instructions for preparation and use, which vary for these types, in cases where specific dilution instructions are not provided by polymer suppliers.

Liquid and emulsion polymers:

1. Add 200 to 500 mL of distilled water to a clean 1-L volumetric flask.

2. After shaking the product container vigorously, weigh out 0.20 g of the polymer product onto an aluminum or plastic weighing dish.

3. Using a distilled water squeeze bottle, rinse all of the polymer into the volumetric flask.

4. Fill the volumetric flask to the 1-L mark with distilled water.

5. Cap and shake for at least 1 min.

6. The strength of the stock solution will be 200 mg/L or 0.20 mg/mL.

7. Therefore, 1.0 mL of the stock solution added to a 2-L jar will be equivalent to a dose of 0.10 mg/L.

8. Most polymer stock solutions of this strength will degrade within 24 hr.

In order to compare the performance of a liquid polymer product to that of a dry product, the polymer content of the liquid should be obtained from the manufacturer. Many liquid polymer products contain between 10 and 40 percent of active polymer. Therefore, for a 20 percent solution, a dosage of 1.0 mg/mL of liquid polymer product contains only 0.20 mg/L of actual polymer. Dry polymers are essentially 100 percent polymer.

Dry polymers:

1. Add 200 to 500 mL of distilled water to a clean volumetric flask.

2. Drop in a magnetic stir bar and place the flask on a magnetic stirrer.

3. Weigh out 0.20 g of the polymer onto an aluminum or plastic weighing dish.

4. Using a distilled water squeeze bottle, rinse all of the polymer into the volumetric flask.

5. Mix at medium speed for at least 2 hr.

6. Remove the stir bar and fill the volumetric flask to the l-L mark with distilled water.

7. Cap and shake it for at least 1 min.

8. The strength of the stock solution will be 200 mg/L or 0.20 mg/mL.

9. Therefore, 1.0 mL of the stock solution added to a 2-L jar will be equivalent to a dosage of 0.10 mg/L.

10. Most polymer stock solutions of this strength will degrade within 24 hr.

Polymers are expensive, but they usually work well at low concentrations. The maximum dosage should be less than 2–3 mg/L, and dosages as small as 0.05 mg/L may prove effective as aids for coagulation or flocculation. Note the maximum permissible dosage listed on the NSF International certification, and ensure that this is not exceeded.

pH Control. Control of pH during jar testing is important for a number of reasons:

-   Simulating existing conditions

-   Evaluating alternatives for optimum turbidity removal

-   Evaluating alternatives for optimum dissolved organic carbon (DOC)/ trihalomethane (THM) precursor removal

-   Minimizing coagulant residual in the distribution system

If the coagulation reaction using aluminum or iron salts as the primary coagulants occurs at nonoptimized pH conditions, the quality of the treated and filtered water can be degraded by elevated concentrations of dissolved aluminum or iron, as discussed in chapter 1. Particularly when alum is used, the metallic salt coagulants are generally more susceptible than the polymers to loss of effectiveness under nonoptimum pH conditions. Keep in mind that both alum and ferric coagulants will lower the pH of the water, especially when the buffering capacity of the water is low.

Often, pH is controlled only by the dosage of the coagulant applied. If the water treatment plant operates this way, jar tests should also be conducted the same way to find the optimum chemical dosage for routine treatment of water.

Sometimes, however, the properties of raw water cause optimum treatment to occur at a pH significantly different from that obtainable from the coagulant alone. A jar test series will quickly demonstrate this condition. In some cases, cost-effective treatment would adjust the raw water pH with another chemical (acid or base) to achieve the best pH for the coagulant of choice. Again, the jar test is perhaps the most valuable tool for rapidly determining the best combination of coagulants and other chemicals to achieve the most cost-effective and process-efficient reaction pH.

Adjusting pH during a jar test can be a hectic task involving measurements of and adjustments to pH in six jars during a 1-min rapid mix. The easiest and generally most accurate method is to predetermine the required acid or base dose by conducting a titration. This is done by taking a small sample of raw water, say 100 to 200 mL, and adding the coagulant dosage equivalent to that which will be added to the 2-L jar. Place the sample on a standard stirrer and titrate with acid or base, recording the dosage to reach the desired pH level. A single titration can determine the acid or base dosages needed to reach different pH levels. With this information, the proper volume can be premeasured and fed to the jar during, before, or after coagulant addition. The pH at the end of the rapid mix phase still should be measured and recorded. Remember, however, that the pH should be measured both before and after the jar test. In some waters, the phenomenon of pH increase throughout the jar test is caused by CO_2 off-gassing.

Common pH adjusting chemicals.   A number of chemicals are routinely used for pH control including the following:

- Lime

- Sodium hydroxide (caustic)

- Hydrochloric acid

- Sulfuric acid

- Carbon dioxide

- Soda ash

Historically, because water treatment has focused on turbidity removal, lime and caustic have been the most widely used for pH adjustment because metal coagulants react with alkalinity and can depress pH in water with low alkalinity. Furthermore, these chemicals would be used to produce water that is stable with respect to calcium carbonate deposition for distribution system corrosion control. Currently, however, increasingly stringent disinfection requirements in the Surface Water Treatment Rule (SWTR) and regulations for disinfection by-products (DBPs) are causing a shift toward maximizing DOC/DBP precursor removal. Because optimum organics removal tends to occur at relatively low pH values, the use of acid to depress pH is increasing.

Lime.   In many cases, the use of caustic rather than lime to raise pH is more convenient for jar testing. Lime is dosed in a suspension that requires continuous stirring, but caustic is in solution and can be used much more easily. The +2 charged calcium ions associated with lime have been shown to act as a coagulant aid. Consequently, its use for increasing pH may have a slightly more beneficial effect than the use of caustic, as the sodium ion has only a single positive charge. However, in many cases, the difference provides negligible benefits. If a base is required for pH adjustment in a jar test, either is generally acceptable, regardless of what the full-scale facility is using, and caustic is much easier to use for jar tests.

Once the dosage of lime or caustic has been established, the following conversion factors can be used to adjust values if the water treatment plant and the jar test series use different products:

- mg/L CaO = mg/L $CaCO_3$ × 0.56

- mg/L $Ca(OH)_2$ = mg/L $CaCO_3$ × 0.74

- mg/L Na_2CO_3 = mg/L $CaCO_3$ × 1.06

- mg/L NaOH = mg/L $CaCO_3$ × 0.80

Sodium hydroxide (caustic soda): For jar testing, a 0.1N solution of sodium hydroxide is generally sufficient for pH adjustment. Reagent-grade sodium hydroxide is usually supplied in pellet form. To prepare a 0.1N solution:

1. Add 200 to 500 mL of distilled water to a clean 1-L volumetric flask.

2. Drop in magnetic stir bar and place the flask on a magnetic stirrer.

3. Weigh out 4 g of the sodium hydroxide pellets onto a plastic weighing dish.

4. Pour all of the pellets into the volumetric flask. Mix at medium speed until all of the pellets dissolve.

5. Remove the magnetic stir bar and fill the volumetric flask to the 1-L mark with distilled water.

6. Cap and shake it for at least 1 min.

The strength of the stock solution will be 4,000 mg/L, or 4 mg/mL as 100 percent NaOH. Therefore, 1 mL added to a 2-L jar will be equivalent to a dosage of 2 mg/L.

Oxidants and disinfectants. In an effort to control disinfection by-products while simultaneously meeting disinfection requirements, it may be necessary to investigate alternative oxidants and disinfectants. Bench-scale testing can provide a way to screen the effectiveness of alternative oxidants and their effect on disinfectant by-product formation.

Chlorine, chlorine dioxide, ozone, and potassium permanganate or sodium permanganate are the chemicals most commonly used as primary disinfectants and oxidants. Chloramines are considered a weak oxidant and generally not suitable for primary disinfection. Rather, chloramines are most often used as a secondary disinfectant to provide residual disinfection in the distribution system. Advanced oxidation processes include the use of hydrogen peroxide and UV radiation in combination with ozone.

Chlorine. Chlorine is the most common oxidant/disinfectant in the water industry. It is available in liquid or gaseous form in pressurized metal tanks, as a concentrated aqueous solution (sodium hypochlorite), or as a solid (calcium hypochlorite). Whatever form is added, chlorine disproportionates into Cl_2, HOCl (hypochlorous acid), and OCl$^-$ (hypochlorite ion). Addition of liquefied or gaseous chlorine decreases pH and alkalinity, while applying hypochlorite solution increases them. Because of the safety concerns that accompany the use of chlorine gas, sodium hypochlorite solution is recommended for jar testing. Sodium hypochlorite is widely available as common household bleach. If concern about impurities in bleach arises, laboratory grade NaOCl solution should be used. Of course, the same chemical may also be available from the full-scale plant. Although household bleach is generally assumed to be 5 percent by weight active ingredient, this should be verified before use, as described below.

The chlorine content of commercial sodium hypochlorite solutions is often expressed in percent by weight (%w/v). If the specific gravity of the liquid is 1, a percent by weight value can easily be converted into a mass concentration as Cl_2 by multiplying by 10,000. Depending on the concentration of the chemical, an appropriately diluted stock solution may have to be prepared. For example, a laboratory-grade NaOCl solution contains 5 percent Cl_2 by weight. This corresponds to 50,000 mg/L as Cl_2. Add 10 mL of the solution to a 1-L volumetric flask and fill to the 1-L mark using distilled water. The resulting stock solution contains 500 mg or 0.5 mg/L chlorine, therefore 2 mL added to a 2-L test jar equals a 0.5-mg/L dose. This can be checked by using a chlorine residual test kit. Because the solution is not stable over time, the chlorine content of the solution must be verified before each use.

Chlorine dioxide. This disinfectant is generated continuously in treatment plants, because it rapidly decomposes. Problems associated with bench-scale evaluations focus mainly on obtaining a high-quality stock solution. If the chemical cannot be obtained from the treatment plant for immediate use, it is recommended to generate it in the laboratory as outlined in Method 4500 ClO_2 in *Standard Methods* (APHA et al. 2005). When applying the chemical during jar testing, a gas-tight syringe is recommended because of the tendency of the chlorine dioxide to degas from the solution. In addition, the stock solution should be standardized prior to application. Measurements should be made of residual chlorite concentrations in any test of chlorine dioxide, since this disinfection by-product is regulated.

Ozone. This disinfectant also must be generated on site because of its instability. A typical ozonation system consists of a generating unit, where the ozone is produced by electrical discharge in the presence of oxygen, and a contactor where the ozone-containing gas is contacted with the water. The use of ozone is limited in jar testing, for it is difficult to apply it to the jar as a concentrated solution. Therefore, sometimes the raw water is ozonated prior to transfer to the jars using a continuous flow reactor. In some laboratories, a relatively high concentration stock solution (about 40 mg/L ozone) in low pH solution is made, and then this solution is added to the jars.

Potassium or sodium permanganate. This moderately strong oxidant does not cause DBP formation, but it may result in pink water at high concentrations. For jar testing, a permanganate solution can be prepared from the powdered or granular solid. Dissolve 1 g of potassium permanganate ($KMnO_4$) in distilled water and dilute to 1,000 mL. The resulting solution contains 1,000 mg/L or 1 mg/mL potassium permanganate, so adding 1 mL of the stock solution to 2 L of raw water will result in a 0.5 mg/L dosage. When evaluating potassium permanganate as an alternative oxidant, it is recommended to determine the appropriate dosage before beginning jar tests by adding various dosages to beakers containing the raw water and selecting one that does not result in pink water after a preset period of time, e.g., 30 min. Sodium permanganate is available as a liquid, and this compound is sometimes used to avoid the problems associated with feeding dry permanganate.

Chloramines. This weak oxidant forms a persistent residual, making it a suitable secondary disinfectant in the distribution system. Chloramines are formed by the reaction of ammonia with chlorine. Jar tests may be used to evaluate how the point of ammonia application affects disinfectant by-product formation. Although ammonia can be applied in various forms in full-scale treatment, for jar testing it is convenient to make up a solution from ammonium sulfate or ammonium chloride. For a 1,000-mg/L NH_3-N solution, dissolve 3.82 g of anhydrous ammonium chloride or 4.72 g of anhydrous ammonium sulfate in distilled water and dilute to 1 L. Because some DBPs are quite volatile (for example chloroform, one of the regulated THMs), care must be taken to minimize volatilization during the jar test. This is sometimes accomplished by allowing a sheet of bubble wrap to float on the water surface during the test.

Advanced oxidation processes (AOPs). Treatment methods such as ozone/UV and ozone/hydrogen peroxide processes can be highly effective alternatives to traditional oxidants. AOPs are commonly evaluated in pilot-scale tests, rather than in bench-scale tests.

Other chemicals. Successful simulation of full-scale plant conditions requires addition of any other chemicals normally added during rapid mix, flocculation, or sedimentation at the same dosages in the jar tests. These additional chemicals may have profound effects on treatment performance. For example, fluoridation chemicals can complex with alum if fluoride is added in pretreatment, resulting in the need to add more alum to attain proper coagulation. Some phosphate chemicals that are used for corrosion control can cause increased filtered water particle counts and turbidity when filters are returned to service if filters are backwashed with water containing these chemicals (Amburgey et al. 2004).

If testing includes laboratory analyses that cannot be conducted immediately, samples must be preserved according to standard analytical practices. Preparations for a jar test must ensure that the appropriate preservatives are on hand. Commonly used preservatives may include nitric acid for metal analyses, sodium thiosulfate for quenching DBP samples, or ethylenediamine for preserving by-products of chlorine dioxide disinfection. Refer to documentation for the appropriate method or consult the laboratory conducting the analyses.

Data Collection and Documentation

It is critical that the data from the jar test be recorded. Unfortunately, sometimes operators run a jar test and do not fill in the jar test data sheet. When this happens, the results of the test are only useful for that operator at that time. By recording all of the information, the results can be used in the future when a similar set of water quality conditions arise.

During jar testing, several events may occur simultaneously, and the actual evaluation of the test results may not be made until several days after the test are performed. Significant observations may be difficult to recall if not written down. This documentation requires preparation of data sheets and test protocols, which serve not only for data collection but also as reminders for the steps to take during a jar test. The data sheets should therefore be designed so that the form holds all relevant information for the jar test in the sequence that the data are collected.

Figure 2-15 shows an example data sheet. In preparing for the jar test, use of such a data sheet allows entry of source water quality parameters, concentrations of the chemicals, and the hydraulic characteristics of the full-scale plant and corresponding jar test parameters. Do not underestimate the importance of visual observations of a trained operator; for example, the relative speed with which flocs form the jars (e.g., jar 3 formed visible floc before jar 1) and the appearance of the jars (e.g., milky, cloudy, good phase separation, large flocs, etc.). Space must be available on the form for noting such visual observations.

It is best to prepare these data sheets using computer spreadsheet programs. These tools allow both rapid calculation of removal efficiencies and capabilities for graphical representation of results.

As with any analytical work, repetition is important to assess the amount of variation and enhance quality of data. Whenever possible, tests should be run more than once, and data compared.

CONDUCTING THE JAR TEST

After the preparation tasks described in the previous section have been performed, the actual jar test can be conducted. At this point, the operator should have defined the study goals (e.g., optimizing coagulant or polymer dosage) and the testing parameters (hydraulic characteristics of the plant, points of application, etc.). Testing and analytical equipment should be ready (jars, stirrer, properly labeled sample containers, turbidity meter, pipettes, etc.). All reagent solutions should be prepared (coagulants, polymers, oxidants, etc.), and a data sheet should be available in a convenient spot.

The operator should ensure that all chemicals are properly labeled and that the reagent solutions are thoroughly mixed. The reagent containers should be placed near the jar test equipment in the order that they are used. Pipettes and syringes should be labeled, too, and placed in or next to the corresponding reagent containers. Automatic pipettes should be set to the correct volumes.

The jars and the paddles of the stirring mechanism should be cleaned by wiping with a damp cloth and rinsing with warm tap water to remove any residue from previous jar tests.

The data sheet should be a good guide for conducting the jar test:

1. Treatment performance data are often expressed in terms of percentage removal. Therefore an important beginning step is to determine the quality of

Date:		Time:
		Concentration (mg/L)
Coagulant:		
Oxidant		
Polymer:		

Source Water				
pH	Turbidity (ntu)	Alkalinity (mg/L as $CaCO_3$)	UV254 (cm^{-1})	TOC (mg/L)

Jar Number			1	2	3	4	5	6
Rapid Mix		G (s^{-1})						
		rpm						
		Duration (s)						
Flocculation		G (s^{-1})						
		rpm						
		Duration (s)						
Coagulant Dose (mg/L)								
Volume of Coagulant Added (mg/L)								
Oxidant Dose (mg/L)								
Volume of Oxidant Added (mL)								
Polymer Dose								
Volume of Polymer Added (mL)								
Coagulation pH								
Settling Velocity (cm/min)	Depth of Sampling (cm)	Time of Settling (min)	Turbidity (ntu)	Turbidity (ntu)	Turbidity (ntu)	Turbidity (ntu)	Turbidity (ntu)	Turbidity (ntu)
TOC (mg/L)								
UV-254 (cm^{-1})								

Figure 2-15 Example jar testing data sheet

the raw water to be tested. Such data may be obtained from treatment plant records or determined during the jar test. Also, pH and alkalinity data may help determine necessary additions of acid or base.

2. Enter the names and concentrations of the chemicals to be added on the data sheet. This information is necessary to determine the volumes of chemical to be added during the jar test.

3. Enter the G values for the rapid mix and flocculation stages of the full-scale plant on the data sheet. If the effect of varying mixing intensities is to be evaluated, use the appropriate range of G values. Convert these values to the appropriate rpm in the jar test. Depending on the jar test equipment used, refer to Figures 2-8 through 2-10 to determine the correct rpm value.

4. Enter the detention times for the rapid mix and the flocculation stages of the full-scale plant in the data sheet. If the effects of detention times are to be evaluated, use an appropriate range of durations.

5. Enter the coagulant dosages on the data sheet. If the test will determine an optimum coagulant dosage, it is useful to select dosages in increments of 5 to 10 mg/L for alum, or equivalent dosages if other coagulants are to be tested. Smaller increments may be used for fine-tuning the optimum coagulant dosage. Then calculate the volume of coagulant to be dispensed into the jar to obtain the desired coagulant dosage.

6. If applicable, proceed with all the other chemicals in a similar manner.

7. Based on the surface loading or overflow rate of the sedimentation basin of the full-scale plant, determine the critical settling velocity as outlined above. Divide the depth of the sampling port on the jars in centimeters by the settling velocity to obtain the sampling time in minutes.

8. Fill the jars with the water to be tested, and position them under the stirring apparatus so they are centered with respect to the impeller shafts.

9. Lower the impellers or paddles so that they are about one-third from the bottoms of the jars.

10. Begin the flash mix period based on the previously determined values. Do not forget to record the starting time. Dispense the desired quantities of chemicals as rapidly as possible into the jars. Dispense the chemicals in the same sequence as at the full-scale plant, unless the effect of moving the point of application is to be evaluated.

11. After the rapid mix period, decrease the mixing speed to the predetermined value for the flocculation period. At this point, the coagulation pH is typically measured.

12. After the flocculation period, stop the mixer and remove the paddles from the jars. Collect samples at the times previously calculated to simulate the full-scale sedimentation basin. Sample withdrawal may be accomplished either by the use of a syringe, a fixed sampling port, or a pipette. The first portion of sample taken from a fixed port should be discarded. When a syringe is used, samples should be taken from the same depth as the fixed port.

13. After sampling, conduct the laboratory analyses, observing holding times required for specific analytical applications.

14. Enter laboratory results on the data sheet.

Standard Operating Procedures

As with other routine tasks, it is a good idea to develop a detailed standard operating procedure (SOP) for the jar test that is specific to a particular plant. The SOP helps ensure that all operators conduct the test in the same manner and also serves as a training tool. Example SOPs used by water agencies for conducting jar tests are in the appendix of this manual.

INTERPRETING AND PRESENTING THE RESULTS _____

Successful interpretation of jar test results is possible only when adequate data have been collected and recorded. Comparisons of jar test data require such key information as all water quality data, hydraulic data, and the types and dosages of the chemicals used. Once all the data are collected, the easiest way to evaluate them is to prepare charts and graphs using computer spreadsheet programs. Well-prepared charts and graphs show how well the jar tests simulate the full-scale plant as well as the effects of alternative treatment options.

Guidelines for Charts

Percentage removal versus absolute values. Jar test data can be easily converted into percentage removal values using the formula:

$$\text{Percentage removal} = (1 - \frac{\text{Final value}}{\text{Initial value}}) \times 100$$

This method may be chosen to assess treatment performance when initial water quality parameters vary or no absolute concentration is specified as the treatment goal.

Use of percentage removal data may obscure actual differences, so the initial values or ranges of values should be included in the data report. For example, consider the following data:

Description	Jar Test 1	Jar Test 2
Source water turbidity (ntu)	18.0	7.0
Settled water turbidity (ntu)	2.2	1.0
Percentage removal	87.8	85.7

Although the percentage removal is higher in jar test 1, the hypothetical treatment goal of 1.0 ntu in the settled water is reached only in jar test 2.

Logarithmic scales. When the data to be displayed in a set of charts include values in increasing increments, such as 0.05, 0.2, 0.5, 5, 20, 100, a logarithmic scale may prove helpful to compress the distances between the data points.

Smoothing functions. Spreadsheet programs can display data by connecting the data points with a smooth curve. However, the operator must keep in mind that these curves do not represent the actual data, so they may lead to misinterpretation of the results. It is recommended that the operator plot the actual data points and use straight lines to connect the points to improve readability.

Trendlines. When using a computer to graph large amounts of jar test data, the program may allow display of a trendline or "best-fit" line. This option calls for a careful choice of the correct statistical method for calculating the trendline to avoid a misleading result. In many cases, a trendline can be well approximated by just "eyeballing" the data points. The use of trendlines is especially helpful when many data points are displayed in the same plot. Sometimes, however, a trendline may be misleading. In the case of a plot of settled water turbidity on the y-axis and increasing coagulant dose on the x-axis, the results are often not a straight line. The use of a trendline through the data may obscure a "dip" in the trend, which may indicate optimal performance.

Error bars. If the precision of the analytical method is known, instead of just the data point, error bars can be included in the graph. This addition allows an assessment of whether two data points are significantly different from one another.

Types of Charts

Several methods are available for reporting or reviewing the results of jar testing. The graph type selected depends on the preference of the operator and the intended use. A bar graph is suitable only for comparing a few values in different categories in a single plot. Figure 2-16 represents a comparison of jar testing results with the results from a full-scale plant. Note that the percentage removal values for turbidity, UV absorbance, DOC, and THM formation potential (THMFP) were comparable between the full-scale plant and the jar tests. In addition, similar amounts of THMs were formed as a result of prechlorination. These results indicate that the jar testing protocol successfully simulated the full-scale treatment plant. The bar graph in Figure 2-17 demonstrates how jar test results may aid in selecting a suitable synthetic organic polymer for enhanced

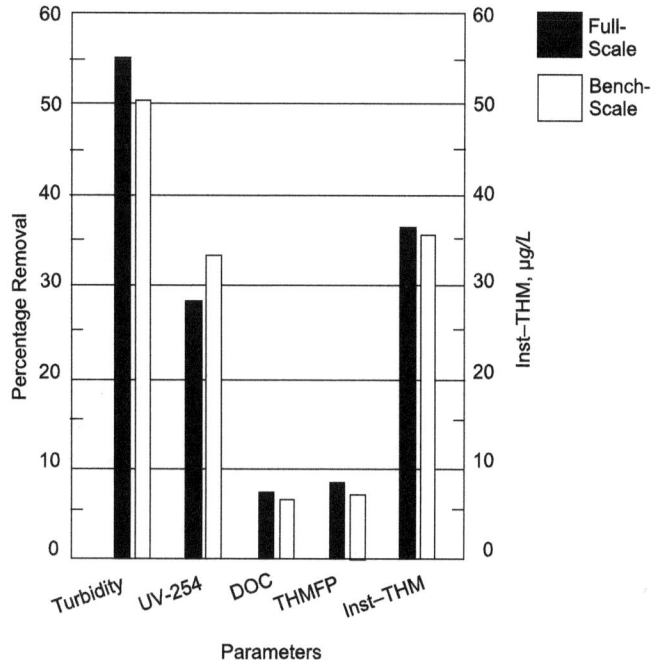

Source: City of Phoenix, Ariz. 1989.

Figure 2-16   Example correlation between jar test results and full-scale plant performance

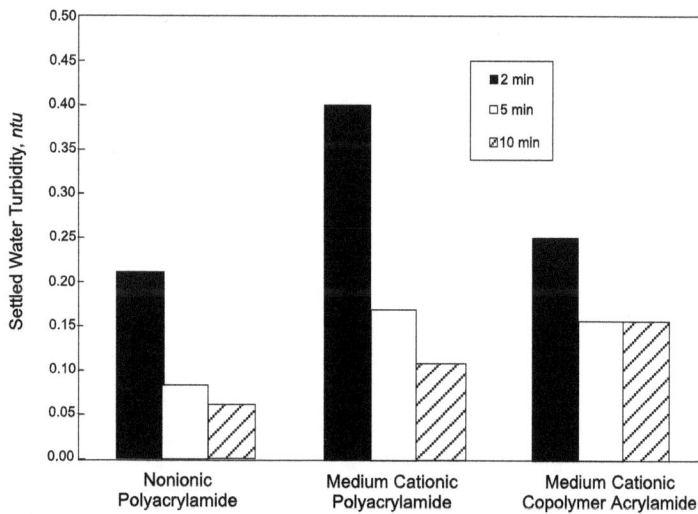

Source: AH Environmental Consultants Inc. 1997.

Figure 2-17   Use of the jar test to select coagulant aids: Turbidity versus settling time

solids–liquid separation. The chart indicates that nonionic polyacrylamide performed best in the jar tests at all three sampling times, i.e., overflow rates.

An *x–y* (scatter plot) graph is suitable for displaying a series of data points. Multiple lines can be used to express results under various conditions. For example, in Figure 2-18, UV-absorbance data were collected from six jars that received increasing alum dosages. The test was then repeated at a different pH. This graph shows UV-254 absorbance versus alum dosage, and each line represents the data series for one pH

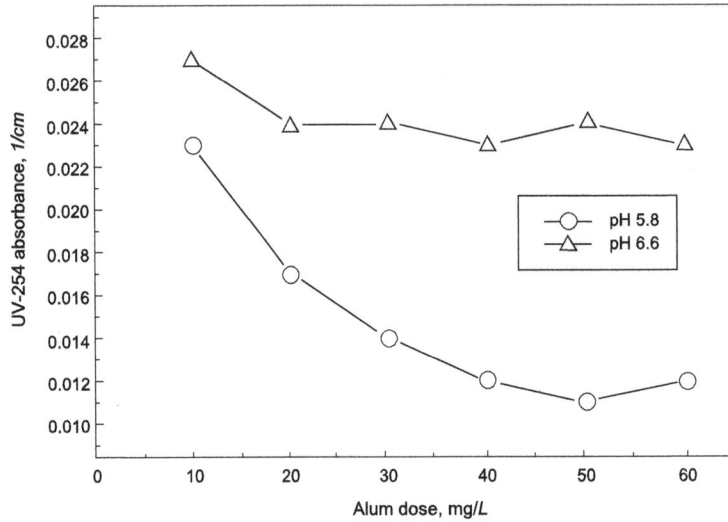

Source: AH Environmental Consultants Inc. 1997.

Figure 2-18   Use of the jar test to optimize the coagulation pH: UV-254 versus alum dose

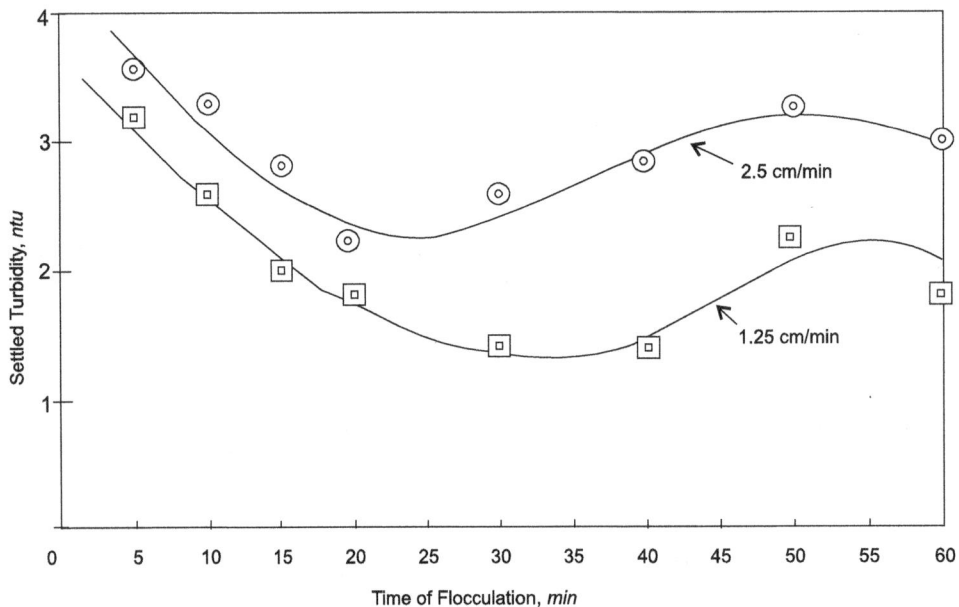

Courtesy of J. Edward Singley and Tim Brodeur.

Figure 2-19   Example use of the jar test: Flocculation time versus settled turbidity

level. The plot indicates that coagulation at pH 5.8 yielded lower UV-254 absorbance than coagulation at pH 6.6. Another example of an x–y graph, shown in Figure 2-19, illustrates the effect of flocculation time on settled water turbidity at two different settling velocities, i.e., overflow rates. The graph in Figure 2-20 was used to determine the optimum polymer dose for a treatment plant based on jar tests. The logarithmic scale on the y-axis helps illustrate the differences in turbidity.

Topographs can be used to simulate a three-dimensional view of the collected data. They are commonly used to represent large data sets, and they often require special software packages. Figure 2-21 shows a topograph of THM formation potential as a function of applied ozone and alum dosage. A case study in chapter 7, "Relationships

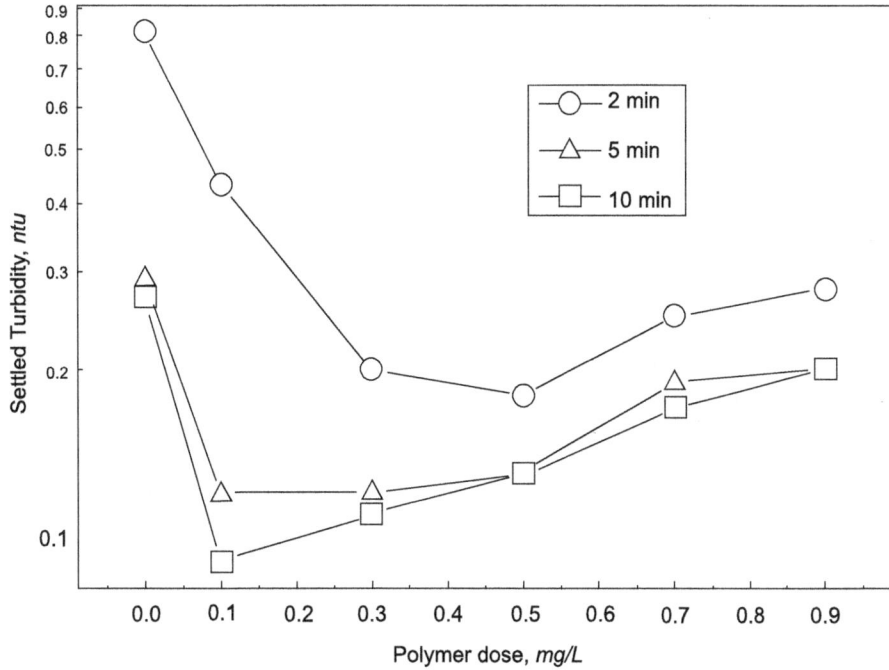

Source: AH Environmental Consultants Inc. 1997.

Figure 2-20   Use of the jar test to determine the optimum polymer dose: Turbidity versus dose and settling time

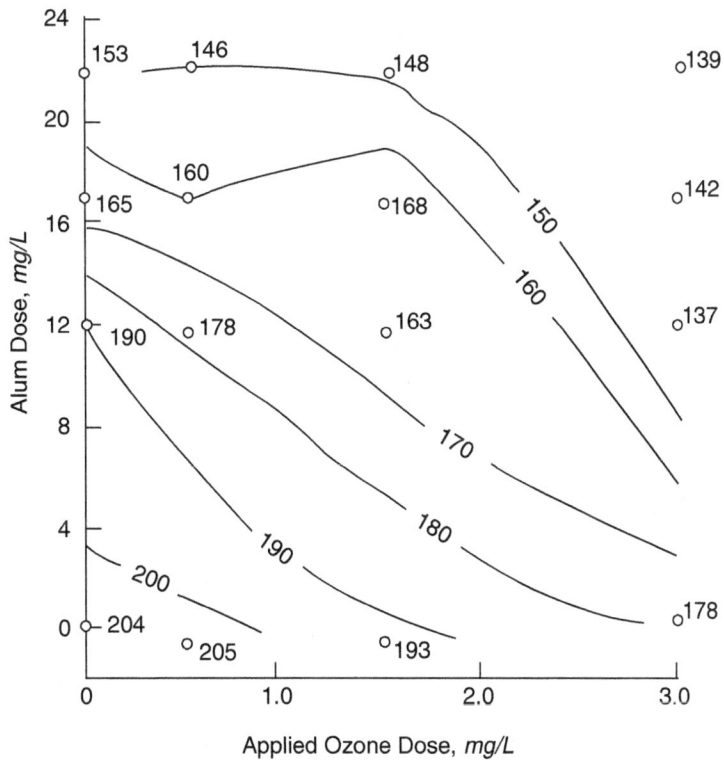

Source: Environmental Engineering & Technology Inc. 1987.

Figure 2-21   Alum dose, ozone dose, THMFP topograph

Between Coagulation Parameters, Winston-Salem, N.C.," utilizes topographs and other types of charts as aids for interpreting data developed in an extensive program of jar testing.

SPECIAL APPLICATIONS

Bench-Scale Evaluation of Filtration

The purpose of a filter is to remove particles from the water that flows through it. Current theories indicate that effective particle removal in filtration depends on both physical and chemical factors. Effective filter performance depends not only on the standard design criteria of filtration rate, media size, and bed depth but also on

- Transport of the particles to the surfaces of the filter media

- Attachment of the particles to the media surfaces

- Concentration and size of the particles applied to the filters

Therefore, filtration is a difficult water treatment process to evaluate using bench-scale techniques. Consequently, the similarity between laboratory and full-scale filter performance cannot be assured. Nonetheless, bench-scale filtration techniques are available to quantify some of the required filtered water quality information and to establish trends in performance.

Available bench-scale filtration alternatives include

- 0.45-µm membrane filters.

- 1-µm glass fiber filters

- Whatman 40 (8-µm) filters

Choice of Bench-Scale Filter Test Apparatus

The selection of a specific method depends on the information required and the ability of the method to produce filtered water quality similar to that from the full-scale filters. For example, based on standards for 0.45-µm membrane filters, any NOM that passes through the filter can be operationally defined as in the dissolved phase rather than the particulate phase. Consequently, if the objective of the study/testing program is to determine the efficiency of the treatment process at converting the dissolved organic material to the particulate phase, the 0.45-µm filters would be recommended.

For some studies, the objective is to evaluate the effects of various treatment alternatives on filtered water quality. Typical water quality parameters of concern include color, chlorine demand, TOC, and disinfection by-product formation potential (DBPFP). The 1-µm glass fiber filter is usually sufficient to meet these objectives.

Whatman 40 (8-µm) filters are most often used for comparing filtering effectiveness, such as comparing the use of filter aids or conducting bench-scale direct-filtration studies. However, filter performance data such as head loss accumulation rate and filter run time collected in this way may be misleading.

Using Water From Rapid Mix

This procedure is applicable for evaluating water after coagulants such as alum or ferric have been added but before polymer addition. Water from the full-scale plant's rapid mix can be used when a coagulant or flocculent aid such as a polymer is to be investigated, without modifying the alum or ferric dose. The procedure is similar to the one given previously but is easier to perform, as the water to be tested has already been dosed with one or more chemicals during rapid mix. All steps are the same, except that coagulant addition and rapid mixing are omitted, so the jar test begins with polymer addition, if this is being evaluated, and then flocculation. If chemicals for pH control have been added in the plant prior to the sampling point, these steps are also omitted. In order to be successful, the time in between collecting the coagulated water and the running of the jar test must be very short (not more than a few minutes). Water from the plant's rapid mix might also be used in cases where variations in flocculation speeds or times are being investigated.

Decreasing the Primary Coagulant Use With Coagulant Aids

In certain situations, it may be possible to decrease the dosage of primary coagulant and achieve the same level of performance when using a small amount of a coagulant aid. The test differs from the preceding procedures in that the coagulant dosage is varied, while the previously selected coagulant aid dosage is held constant. The range to be evaluated will be determined by the dosage presently used in the plant. For example, in a jar test to evaluate a current dosage of 100 mg/L alum under highly turbid conditions, a good range to try might be 10, 30, 40, 60, 80, and 100 mg/L alum. In this example, the first jar would be dosed with 10 mg/L, with increasing dosages in subsequent jars until 100 mg/L is reached. The polymer dosage being tested would be same in each jar.

If lime or another pH-controlling chemical is presently employed in the plant, this dosage must be adjusted since its addition is meant to counteract pH changes caused by coagulant addition. The amount added to achieve a particular pH should be adjusted to be proportional to the coagulant dosage. The titration procedure discussed earlier will provide the needed data.

In addition, the final pH of each jar test should be measured. If these pH values are not within 0.5 pH units of the settled water in the plant, readjust the lime dosages and repeat the jar test. A more time-consuming procedure can be used to locate the precise combination of optimal coagulant and coagulant aid dosages. An entire set of coagulant aid additions can be evaluated at each primary coagulant dosage, thus covering many combinations. For example, six polymer dosages (perhaps 0, 0.1, 0.25, 0.5, 1.0, and 1.5 mg/L) could be employed in each jar test set, while maintaining a constant alum dosage. The procedure would then be repeated using alum dosages (in each of the six jars) of 10, 25, 50, 75, and 100 mg/L alum. The optimal dosage combination can then be determined. The disadvantage of this approach is obviously the large number of jar tests that must be run; note that a very large raw water sample should be obtained so that all of these tests are run on the same water. This precaution is particularly important if coagulation is being evaluated under storm runoff or other rapidly changing conditions.

Nonconventional Treatment

Attempts are sometimes made to apply jar testing procedures to nonconventional treatment processes such as upflow clarifiers, sludge blanket clarifiers, and softening processes. No jar testing procedures have been established for nonconventional settling basins such as upflow clarifiers and sludge blanket clarifiers. Some manufacturers have applied empirical methods to develop procedures for evaluating their process units, but some of these procedures may be of questionable value.

One manufacturer of an upflow sludge blanket clarifier recommends that a "stickiness coefficient" be calculated with the aid of a special test apparatus to which "jigs" of chemically treated water are added to simulate the pulsing of the sludge blanket. The applicability of this type of bench-scale procedure to full-scale treatment units should be verified before implementation.

Some nonconventional processes allow reasonable empirically established modifications to the standard bench testing procedures. For example, one bench-scale simulation of a lime softening process including a solids contact unit required the addition of 2.5 g/L of calcium carbonate to the jars during the flocculation step in order to generate results similar to those from the full-scale facility.

Jar Tests for Direct Filtration Plants

Direct filtration systems usually have raw water that will be processed through a mixing step prior to flowing onto the filters. There is no sedimentation step, and sometimes no distinct flocculation step, at these plants. The process relies primarily on the filters, which are often deep-bed media filters. Therefore, a jar test consisting of rapid mixing followed by a filter index test might be a useful tool in assessing the chemistry, dosages, and mixing requirements that will provide the optimal particle removal.

Conducting the direct filtration jar test. Fill in the data sheet with the appropriate information regarding raw water quality, chemicals being tested and their dosages, order of addition, as well as mixing speed and time before collecting samples.

1. Fill sample jars with raw water and position each of them under the stirring apparatus so they are centered with respect to the impeller shafts.

2. Lower the mixing paddles into the jars so they are about one-third from the bottom of the jars.

3. Start mixer for rapid mix to match the predetermined values for actual plant velocity gradient.

4. Inject chemical into the jars as rapidly as possible and in the same sequence as desired for plant feed locations. Do not forget to start the timer.

5. Match the plant time for rapid mixing.

6. Reduce mixing speed to desired G value for flocculation (if applicable) and run for the appropriate time. If tapered flocculation is employed in the plant, match both the G values and times for each stage of flocculation, or use different G values if the testing is being done to explore effects of flocculation energy on direct filtration.

7. Once the flocculation time has expired, stop mixers, remove the paddles, and collect sample for the Filter Index Test.

Filter Index Test. The purpose of this test is to determine the filterability of the treated water resulting from the mixing process. The test involves measuring the time required to filter a known quantity of water through a micropore filter and comparing it to the time required to filter distilled water. The particles that are properly treated with the correct chemistry and mixing will not plug up the filter and therefore will allow the water to pass through quickly. Both the time and filtered turbidity are measured to determine the optimal treatment scheme. The apparatus is shown in Figure 2-22 and involves the use of a Millipore filter, filter paper, and vacuum pump.

Conducting the Filter Index Test. On the jar test data sheet, place additional columns to record the time to filter 200 mL of sample from the jar test and the resulting filtered turbidity. The same size filter media and vacuum must be used for all tests to compare treatment schemes.

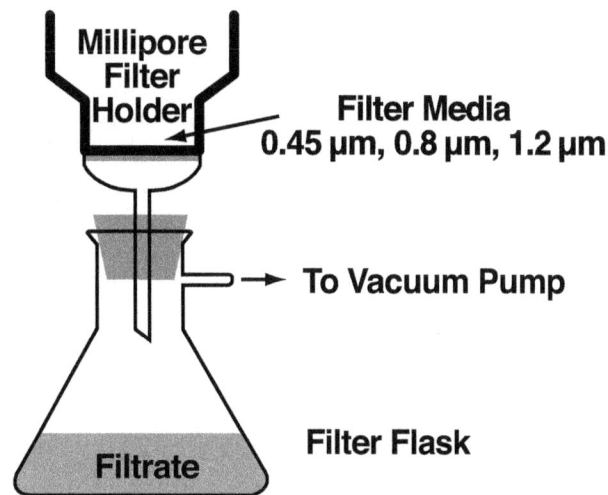

Source: ITT Water & Wastewater

Figure 2-22   Filter Index Test apparatus

1. Measure 200 mL of deionized (DI) water and pour it into the Millipore® filter. Turn on the vacuum pump and record the time it takes to filter the entire volume through the filter paper and the filtered turbidity.

2. Immediately after stopping the mixing sequence in the jar test, collect 200-mL samples from each jar.

3. For each sample, replace the filter paper and record the time to filter 200 mL and measure the resulting filtered turbidity.

4. Calculate the Filter Index by dividing the sample filter time by the DI filter time.

$$\text{Filter index} = \frac{\text{Time to filter sample}}{\text{Time to filter DI water}}$$

5. The comparison of the sample's Filter Index number and filtered water turbidity will result in the optimum treatment scheme for the number of sample schemes evaluated.

6. For additional tests, the treatment scheme that yielded optimum results from all of the prior tests should be the treatment scheme in the first jar, to serve as a benchmark for comparing results of new schemes being tested in the rest of the jars.

The best treatment scheme will provide water that has the lowest Filter Index number (F.I.) as well as the lowest filtered water turbidity. The water sample with low filtered turbidity and high F.I. indicates that the particles in the water were not properly conditioned and the plant would experience filter plugging and short filter runs. The water sample that has high turbidity and low F.I. indicates that the particles are small enough to pass through the filter, not coagulated and conditioned properly to be captured in the filter, and would cause high turbidity in the finished water.

Jar Tests for Softening Plants

The general concepts applicable to jar tests have been described previously in this chapter. For lime softening plants, some minor differences may apply. This section discusses various lime softening treatment trains, reasons for using jar tests at softening plants, and specific considerations for jar testing at lime softening plants. Lime softening chemistry was discussed by Pizzi (1995), Benefield and Morgan (1999), and Horsley et al. (2005) and is not included in this section on jar tests. Treatment trains used in softening include single-stage treatment, two-stage treatment for excess lime softening, and split treatment. A discussion of lime softening process trains is presented in chapter 11 of *Water Treatment Plant Design*, fourth edition (Horsley et al. 2005) and summarized in the following paragraphs.

At high enough pH values, soluble calcium becomes a precipitate and can be removed via settling and filtration. In single-stage treatment, lime is added to raise pH and remove calcium in the form of carbonate hardness, or lime and soda ash are added to remove calcium present as carbonate and noncarbonate hardness. The optimum pH for minimal calcium carbonate solubility is about 10.3. A coagulant may be added to improve clarification. Recarbonation by addition of carbon dioxide follows softening to stabilize the quality of softened water and inhibit precipitation of calcium carbonate onto filter media or water main walls.

Two-stage treatment with excess lime is employed to remove both calcium and magnesium. Mixing and sedimentation are used in each stage. To cause more rapid precipitation of magnesium hydroxide, lime is added in the first stage of treatment to raise the pH to about 11.0–11.3, which is higher than the theoretical value needed to precipitate magnesium. In the second stage of treatment, carbon dioxide is added to lower pH and attain additional calcium removal. A coagulant is added to improve clarification, and if carbonate alkalinity is insufficient, soda ash is used.

Split treatment generally is used for treating groundwater. This process involves bypassing a portion, perhaps 10 to 30 percent, of raw or pretreated water around the first stage of lime softening and introducing it in the second stage of softening. The entire dosage of lime needed to soften all of the water is introduced at the first stage of treatment, creating an "excess lime" condition in that stage. The CO_2 and carbonate alkalinity in the groundwater react with excess lime and cause additional calcium precipitation, and the CO_2 can lower the pH of the softened water somewhat. Split treatment can lower chemical requirements for lime and for carbon dioxide.

Other variations in process trains at lime softening plants include coagulation, flocculation, and sedimentation prior to lime softening; and also include coagulation, flocculation, and sedimentation following single-stage lime softening.

Jar tests at lime softening plants may be undertaken for a variety of reasons, including:

- Verifying theoretical calculations of chemical dosages needed for softening

- Evaluation of clarification after softening

- Evaluation of changes in process operation or of alternative coagulants for improved treatment

- Evaluation of treatment when source water quality has changed

As always, the purpose of the jar test needs to be identified before a jar test program is undertaken.

Multiple stages of treatment and various treatment sequences are employed at lime softening plants, so a single procedure for performing jar tests cannot be prescribed. Instead, the jar test approach needs to be tailored to the plant in a way that mimics the full-scale process to the extent feasible. For example:

- If softening sludge is recycled back to the rapid mix in a conventional mixing/flocculation/sedimentation treatment train used for lime softening, consider adding an appropriate percentage of settled lime softening sludge to the jar test in the rapid mixing step.

- For split treatment, reserve some raw or pretreated water to add in the second stage of lime softening.

- For two-stage treatment, settle and decant clarified water from the jars treated in the first stage and perform the second stage of treatment with the clarified water.

- If clarification is done first, consider use of clarified water from the treatment plant (lime softening basin influent water) if the jar test is performed to evaluate lime softening.

- If the clarification step follows lime softening in the plant, use softened water from the plant for jar tests involving clarification of the softened water.

- Use of chemicals employed at the treatment plant instead of reagent-grade chemicals is recommended for jar tests related to plant operations. To account for the impurities in commercial lime being used at the plant, either obtain lime slurry of a known strength from the plant or prepare a fresh slurry using the lime used for full-scale treatment. This will enable jar test results to relate to lime concentration that would be applied at full-scale.

- To dose lime slurry in tests with multiple jars, remove measured amounts of stirred lime slurry from the stock slurry using a pipette or syringe with a tip large enough that slurry particles will not block it. Dose the measured amounts into small beakers (one for each jar). To add the correct volume of slurry to a beaker, draw the exact volume needed into the pipette. After draining the slurry from the pipette into the beaker, rinse the pipette with lab-grade water, adding the rinsings into the beaker. After all beakers have the appropriate slurry quantities, add lime slurry to all jars simultaneously, and then rinse remaining slurry from each beaker into its respective jar. Lime residue left in the pipette or in a beaker would result in incorrect calculations of lime dosages used in the jar test.

- When chemicals other than lime are added in lime softening at the plant, add them in the same sequence in the jar test.

- To evaluate recarbonation for pH adjustment, acid could be added to depress the pH to the level attained by adding carbon dioxide in the plant.

- If jar tests are done for which chemical reaction time is important, residence time in the jars may need to mimic residence time in full-scale plant processes. This is in contrast to jar tests to evaluate settling of floc, in which short residence times are used for a sampling plan, to reflect a range of particle settling velocities that would be of interest in a full-scale settling basin.

Jar Tests for Ballasted Flocculation Processes

Some high-rate clarification processes use microsand as a seed for floc formation (e.g., the Actiflo® process by Kruger). The microsand provides surface area that enhances flocculation and acts as a ballast or weight. The resulting sand-ballasted floc settles quickly, which allows for clarifier designs with very high overflow rates and short retention times. The sand is then separated from the sludge and recycled. This type of clarification process cannot be mimicked with a standard jar test procedure.

Conducting the ballasted flocculation jar test. A specific jar test procedure has been developed for the sand-ballasted flocculation process, as follows:

1. Prepare several dry polymer solutions at 0.1 percent concentration—cationic, nonionic, and anionic as follows: very slowly, add 500 mg of dry polymer to 500 mL of water in a flask under vigorous mixing with a magnetic stirrer; continue mixing for another 30 to 60 min until all of the dry polymer is dissolved. A new polymer solution should be mixed every 4 to 6 hr.

2. Since jar testing with microsand is very rapid (approximately 5 min/test), it is recommended to run only one jar at a time. Therefore, in order to ensure consistent raw water quality, a large container should be filled with raw water. This container should be stirred each time before a sample is taken for jar testing. In this way, testing can be performed while avoiding any inconsistencies that can be caused by varying raw water characteristics.

3. For best results, test parameters in the following sequence:

   a. At optimum coagulant dose and pH (from the previous experience), check the different polymers at 0.50 mg/L polymer to find which polymer works best.

   b. With best polymer at 0.50 mg/L and optimum coagulant dose, try different pH conditions to identify the optimum pH for coagulation.

   c. With best polymer at 0.50 mg/L and optimum pH of coagulation, try different coagulant dosages to fine-tune coagulant dose.

   d. At optimum coagulant dose, pH, and polymer type, try various polymer dosages (i.e., 0.40, 0.30, 0.20, 0.10, and 0.05 mg/L).

4. Fill a jar test beaker with raw water.

5. Stir the sample at maximum speed and add microsand at a concentration of 10 g/L.

6. At time = 0, add the coagulant and mix at maximum speed for 2 min. If pH adjustment is needed, add alkali before coagulant addition. (Note: This step can be increased to 3 min if necessary for better results. This will depend on the raw water quality.)

7. At time = 120 sec, add polymer and mix at maximum speed for 15 sec.

8. At time = 135 sec, reduce the mixing intensity for 45 sec so that there is just enough energy to keep the microsand in suspension. This is a critical step in testing with microsand. There needs to be enough mixing to keep the sand in suspension, but too much mixing will damage the microsand-ballasted floc.

9. At time = 180 sec, stop all mixing and allow the floc the settle for 2 min.

10. Sample settled for water turbidity, pH, and any other parameters of concern.

11. Repeat procedure for all dosages and parameters.

Jar Tests for Dissolved Air Flotation (DAF) Plants

In the DAF process, raw water particles are flocculated and separated out of the water by floating them to the surface, rather than settling them to the bottom of a basin. The process uses micro-sized air bubbles that form at the bottom of the contactor where they mix with the flocculated solids and float the floc. The air bubbles are produced by recycling a portion of the effluent through a pressurized, packed saturator tank where air is introduced and the water super-saturated with respect to atmospheric pressure. When the supersaturated water is returned to the reaction zone of the flotation tank, microscopic bubbles form and rise to the water surface, carrying up the floc. The floated floc is removed from the top of the basin by mechanical or hydraulic means, while the clarified water is removed by laterals from the bottom of the basin.

The particle size for removal in flotation can be tens of microns rather than the hundreds of microns size required for sedimentation, so both the flocculation and clarification detention times are less than conventional settling processes. The DAF flocculation and clarification basins are a quarter to one-tenth of the size of conventional sedimentation systems. The mixing times are generally rapid mix for 30 to 60 sec followed by 5 to 10 min in each of two flocculation basins with tapered velocity gradient mixing. The surface loading rates of the solids separation part of this high rate process can be from 6 to 20 gpm/ft^2 (15 to 19 m/hr). DAF will produce sludge solids of 2 to 5 percent in the floated floc.

In order to conduct the flotation jar test, the jar test apparatus must include a pressure tank to create the air-saturated water that will mix with flocculated solids in the jars so as to simulate the reaction zone of the DAF. An example DAF jar tester is shown in Figure 2-23.

As in conducting conventional jar tests, all the preparation tasks described in previous sections must be performed including defining the study goals and determining the testing parameters. Typically, the tests are conducted to determine the optimum chemical(s), order of addition and their dosage(s), optimum mixing time and energy (mixing speed), optimum recycle rate, and expected effluent quality. The testing and analytical equipment should be ready, all reagent solutions prepared, and a data sheet

Source: EC Engineering Inc.

Figure 2-23   Jar test equipment for DAF testing

available in a convenient spot to record the conditions used during the test and the analytical results of the water tested from each jar.

The jars and the paddles of the stirring mechanism should be cleaned by wiping them with a damp cloth and rinsing them with warm water to remove any residual from previous jar tests.

Conducting the flotation jar test. Fill in the data sheet with the appropriate information regarding raw water quality, chemicals being tested and their dosages, mixing speed and time, recycle rate, and flotation time before collecting samples.

1. Fill the saturator tank one-third to one-half full of clean water.

2. Ensure the saturator tank lid is sealed properly.

3. Charge the saturator tank with 80–85 psi of air.

4. Shake the saturator tank for 1 min.

5. Insert tubing into the saturator tank quick-disconnect (hose to connect saturator to jar tester).

6. Hook up recycle line to the sample jars.

7. Run the recycle line into the jar to blow out any air in the recycle line and to calibrate an expected volume of recycle water for each setting.

8. Empty sample jars.

9. Fill sample jars with raw water (1 L) and position each of them under the stirring apparatus so they are centered with respect to the impeller shafts.

10. Lower the mixing paddles into the jars so they are about one-third from the bottom of the jars.

11. Start mixer for rapid mix at 200–300 rpm or whatever the predetermined values are to be.

12. Inject chemical into the jars as rapidly as possible and in the same sequence as desired for plant feed locations. Do not forget to start the timer.

13. Let mix at rapid mix speed for 30 to 60 sec.

14. Reduce mixing speed to desired G value for Stage-1 flocculation (usually 100 rpm) and run for desired time (usually 5–10 min).

15. Reduce mixing speed to desired G value for Stage-2 flocculation (usually 50 rpm) and run for desired time (usually 5–10 min).

16. Shake saturator again for 1 min.

17. Once the flocculation time has expired, stop mixers, remove the paddles, and inject a measured quantity of water supersaturated with air into the jars to simulate recycle in the full-scale process (usually 10 percent recycle by volume).

18. Wait 5 min after the supersaturated water injection and draw a 250-mL sample from the sample port in the beaker (DAF effluent).

19. Run the selected analytical test on the DAF effluent sample and record the results.

Based on the first set of results conducted, vary one component with the next flotation jar test. For example, with optimum chemistry, vary the mixing time or energy and leave other variables the same, or with optimum chemistry, vary the recycle rate and leave all other variables the same, and so on.

REFERENCES

AH Environmental Consultants Inc. 1997. Disinfectant/Disinfection By-Product Rule at the Naval Station Roosevelt Roads, PR. *Final Report to Naval Facilities Engineering Command, Atlantic Division*. Hampton, Va.: US Navy, Atlantic Division.

AH Environmental Consultants Inc. 1998. Evaluation of Alternatives for Enhanced Coagulation and DBP Control for the Goldsboro Water Treatment Plant. *Final Report to Department of Public Utilities, City of Goldsboro, N.C.* Goldsboro, N.C.: City of Goldsboro.

Amburgey, J.E., A. Amirtharajah, B.M. Brouckaert, and N.G. Spivey. 2004. Effect of Washwater Chemistry and Delayed Start on Filter Ripening. *Jour. AWWA*, 96(1):97–110.

American Public Health Association (APHA), American Water Works Association (AWWA), and Water Environment Federation (WEF). 2005. Method 2130 Turbidity. *Standard Methods for the Examination of Water and Wastewater*. 21st ed. Washington, D.C.: APHA.

Benefield, L.D., and J.M. Morgan. 1999. Chemical Precipitation. In *Water Quality and Treatment*, ed. R.D. Letterman. 5th ed. New York: McGraw-Hill.

City of Phoenix. 1989. Water Quality Master Plan. Phoenix, Ariz.: City of Phoenix.

Cornwell, D., and M. Bishop. 1983. Determining Velocity Gradients in Laboratory and Full-Scale Systems. *Jour. AWWA*, 75(9):470–475.

Dentel, S. 1986. *Procedures Manual for Selection of Coagulant, Filtration, and Sludge Conditioning Aids in Water Treatment*. Denver, Colo.:

AWWA and Awwa Research Foundation.

Environmental Engineering & Technology. 1987. *A.B. Jewell Laboratory Treatability Study: City of Tulsa, Okla*. Tulsa, Okla.: City of Tulsa.

Environmental Science & Engineering Inc. 1981. *Water Treatment Plant Process Upgrade and Trihalomethane Reduction Study Final Report*. ESE No. 81-209-200. Arlington, Texas: City of Arlington.

Environmental Science & Engineering Inc. 1982. *Southwest Water Treatment Plant Process Design Study*. ESE No. 82-204-400. Arlington, Texas : City of Arlington.

Horsley, M.B., Doug B. Elder, and Leland L. Harms. 2005. Lime Softening. In *Water Treatment Plant Design*, ed. E.E. Baruth. 4th ed. New York: McGraw-Hill.

Hudson, H. 1981. Jar Testing and Utilization of Jar Test Data. *Water Clarification Processes: Practical Design and Evaluation*. New York: Van Nostrand Reinhold.

Pizzi, N.G. 1995. *Hoover's Water Supply and Treatment*, 12th ed. Bulletin 211, National Lime Association. Dubuque, Iowa: Kendall/Hunt Publishing Co.

NSF International and American National Standards Institute. 2005. Drinking Water Treatment Chemicals: Health Effects. NSF/ANSI 60. Ann Arbor, Mich: NSF International.

US Environmental Protection Agency (USEPA). 1998. *Optimizing Water Treatment Plant Performance Using the Composite Correction Program*. EPA/625/6-91/027. Cincinnati, Ohio: USEPA.

This page intentionally blank.

Chapter **3**

Online Sensors for Monitoring and Controlling Coagulation and Filtration

Robert Bryant, Michael Sadar, and David J. Pernitsky

INTRODUCTION

Online sensors are vitally important for the monitoring and control of coagulation and filtration in the modern water treatment plant. With respect to coagulation and filtration, online sensors are commonly used for monitoring changes in raw water quality, monitoring and control of coagulant feed, clarification and filtration processes, and for regulatory compliance. Table 3-1 summarizes the online sensors typically used for various stages of the coagulation and filtration process.

This chapter is divided into two sections. The first is a discussion of the use of online sensors in coagulation and filtration applications. The second section reviews the technical details of these online sensors. Additional technical details on online sensors as well as further information on the selection, specification, and integration of online monitors into water treatment operations can be found in *Online Monitoring for Drinking Water Utilities* (Hargesheimer et al. 2002).

PROCESS APPLICATIONS OF ONLINE SENSORS

Raw Water Monitoring

Raw water quality, specifically the amount of particulate material, the amount and nature of the natural organic matter (NOM), pH, alkalinity, and temperature have been shown to affect coagulation and filtration processes (Pernitsky and Edwald 2006).

Table 3-1   Typical online sensors used in coagulation and filtration applications

Online Sensor Type	Raw Water Monitoring	Coagulation	Clarification	Filtration
pH	1	1	2	2
Turbidity	1	3	1	1
Particle Counter	2	3	2	1
Ultraviolet (UV) Absorbance	1	1	2	2
Total Organic Carbon (TOC)	1	1	2	2
Streaming Current (SC) Monitor	2	1	3	3
Head Loss (differential pressure)	Not used	Not used	3*	1

Courtesy of David Pernitsky.
1 = online monitoring very useful for process monitoring and control
2 = online monitoring moderately useful for process monitoring and control
3 = online monitoring not commonly used for this process
3* used for contact adsorption (roughing filter) clarifiers

Online raw water quality data can be very important for process control for raw waters that are subject to variations in raw water quality as a result of seasonal changes, precipitation-related runoff events, algal activity, or upstream pollution discharges.

Turbidity and particles.   The amount of suspended particulate matter in the raw water can affect the required coagulant dose, as well as the performance of clarification and filtration processes. Particulates are typically monitored with online turbidity measurements. Particle counters are less frequently used.

Raw water turbidimeters are often used to assist in the dosing of flocculants and coagulant chemicals. If the turbidity in the raw water increases, this is generally an indication that the number of particles is also increasing. In such a case, chemical aids that enhance the removal of particles can be adjusted accordingly, or at the very least, alert the operator that changes in chemical dosage may be required.

Raw water samples tend to have a broad range of turbidity, depending on season and weather events. Thus, an instrument with a wide operation range is necessary, so high-turbidity events can be accurately assessed. Many types of turbidimeters can measure high levels of turbidity. These instruments must be able to accept high flow rates so that particle settling and eventual plugging of the instrument do not occur. Further, most influent waters will have some color and particle absorbance and will therefore require a design that can best minimize these interferences. The two most common types of instruments for raw water applications are the surface scatter turbidimeter and insertion probe instruments. From a regulatory standpoint, online source-water turbidity measurements are not required. However, most regulators consider turbidity to be a key parameter to characterize the source water and to assist the operator in selection of the correct treatment strategy.

Although particle counters have been used successfully to monitor raw water quality, they are most sensitive for waters with low numbers of particles. The most successful applications are those that measure water with turbidity well below 7 ntu (Pernitsky and Meucci 2002). Particle counters are subject to plugging, flow disruptions, and sensing errors when used on higher-turbidity waters. When turbidity exceeds 7 to 10 ntu, turbidimeters are generally more appropriate than particle counters for raw water quality monitoring.

TOC and UV absorbance.   For many surface waters, coagulant doses are controlled by NOM concentration rather than by turbidity (Edzwald 1993, Pernitsky and Edzwald 2006). NOM is typically quantified in water treatment plants by TOC measurements or measurements of the absorbance of ultraviolet light at 254 nm. TOC consists of both particulate organic carbon and dissolved organic carbon (DOC). UV absorbance at 254 nm has been shown to be an excellent surrogate for TOC and disinfection by-product (DBP) precursors in certain waters (Edzwald et al. 1985). The exact relationship between UV absorbance and TOC concentration is unique for each raw water source. However, for a given raw water source, increases in either TOC or UV absorbance indicate increasing NOM concentrations, and therefore increasing coagulant demands.

It is important to note that for some water sources, increases in TOC or UV absorbance can occur without an increase in raw water turbidity. If online or frequent bench-top measurements of NOM concentration are not made, undetected changes in NOM concentration can show up as coagulant over- or under-doses, often at the expense of clarified and filtered water quality. The case study, "NOM Measurements for Coagulation Control," presented in chapter 7, provides an example of the application of UV monitoring as an aid to coagulation optimization.

pH.   pH is a key parameter for understanding the condition of the source water, and online measurements of raw water pH are simple and reliable. Knowledge of raw water pH is important in terms of selecting chemical dosages to achieve the desired coagulation pH. Changes in raw water pH (and/or alkalinity) can affect coagulation pH and coagulation performance. As well, changes in raw water pH can often indicate other water quality changes, for example increased turbidity or natural organic matter loads caused by precipitation events.

Streaming current monitors.   Streaming current monitors measure particle charge, and therefore can be used on raw water to monitor changes in the overall charge of particulates in the water. If the charge in the raw water changes, then the streaming current should trend in the direction of increased charge (either positive or negative). At this point, the coagulant type, dosage, or both can be adjusted to compensate for the change in charge in the water. As will be discussed below, streaming current monitors are most often used after coagulant addition.

Coagulation Process Control

Coagulation process control refers to the proper dosing of coagulants, polymers, and pH adjustment chemicals for the purposes of neutralizing the charge on suspended particles and reacting with dissolved NOM to form floc suitable for downstream clarification and filtration processes. Depending upon the process configuration of the plant, online sensors for coagulation process control may be located immediately downstream of chemical addition or downstream of clarification or filtration.

Turbidity and particles.   The addition of coagulant to a raw water results in changes in particle concentrations because of the formation of floc particles. However, turbidity and particle count measurements of the coagulated water (upstream of clarification) are generally not used for process control. The number of particles produced during coagulation and flocculation is not as important as their settling or filtration characteristics, which are not readily determined from turbidity or particle count data.

TOC and UV absorbance. Coagulants react with dissolved NOM to convert it to a solid phase that can be physically removed by clarification and filtration processes. This is accomplished by the formation of insoluble coagulant-NOM precipitates or the adsorption of NOM onto the surface of floc particles (Pernitsky and Edzwald 2006). The conversion of dissolved NOM to a solid phase happens quickly and is complete prior to the clarification and filtration processes.

Online measurements of TOC or UV absorbance can be very valuable for coagulation process control. For example, reductions in TOC or UV absorbance seen in the full-scale water treatment plant can be compared to those seen during jar tests to verify chemical dose selection. Most importantly, changes in NOM removal through the coagulation process, when detected by online instruments, can provide an early warning to operations staff that coagulant adjustments are necessary. It is often useful to compare TOC or UV absorbance data day-to-day or over the course of a week, depending on the variability of the source water. A reduction in the percent removal of TOC or of UV absorbing substances between the raw and treated water may indicate that a change in the concentration or nature of the raw water NOM has occurred. This information can then be used to increase or decrease the coagulant dose, as required. It should be noted that particulate matter can interfere with TOC and UV absorbance instruments. For this reason, TOC and UV absorbance analyzers are often located downstream of clarification or filtration processes, rather than immediately downstream of coagulant addition.

The US Environmental Protection Agency's (USEPA's) enhanced coagulation strategy requires utilities to achieve certain TOC percentage removals through the treatment process (USEPA 1998b). Online sensors can be used to monitor the TOC of the raw and treated waters to determine TOC removals.

pH. As discussed in chapter 1, the pH at which coagulation occurs is a critical process control parameter, as it affects (1) the surface charge of the particles present in the raw water, (2) the charge of raw water NOM, (3) the charge of dissolved-phase coagulant species, (4) coagulant solubility, and (5) the surface charge of floc particles (Pernitsky and Edzwald 2006). In general, coagulation at or near the pH of minimum coagulant solubility results in favorable coagulation conditions for aluminum-based coagulants and minimizes residual Al concentrations. Ferric coagulants are generally most effective between pH 5.5 and 6.5 for controlling turbidity and NOM removal. Both aluminum and iron-based coagulants have been used, however, in lime softening plants in a pH range of 9 to 10 to coagulate particulate matter discharged from a primary softening basin so it will flocculate, settle, and be filtered successfully. It is important to note that the pH of minimum solubility changes with coagulant type and water temperature, as was shown in Table 1-3.

Streaming current monitors. Streaming current monitors are a very common online instrument for coagulation control. The streaming current reading associated with coagulation conditions that result in good treatment plant performance can be used as an operational set-point. If the streaming current reading changes from this set-point, dosages of the treatment chemicals can be adjusted to bring the particle charge back to the previously determined operational set-point. The effectiveness of streaming current technology is site-specific, but if applied properly, it can be an important tool for coagulation control. The most common location for streaming current monitors is after coagulant addition, as measuring streaming current in this location relates to the capability of coagulated water to be removed effectively in clarification and filtration. If a streaming current instrument is used only to monitor raw water, filtration plant staff must be aware that the raw water streaming current data are irrelevant to the suitability of the water for clarification and filtration. Two case studies describing the use of streaming current detectors are found in chapter 7. They are "Net Charge Equals Positive Change," and "Streaming Current Detector Pilot Study: The Detection of a Ferric Chloride Feed Failure."

Clarification Process Monitoring

Online sensors are commonly used to monitor the performance of clarification processes such as sedimentation and dissolved air flotation (DAF). Clarification processes are intended for the physical separation of floc particles; therefore physical parameters such as turbidity are of most use for process control.

Turbidity and particles. Turbidity or particle counting can be used to monitor the effluent of clarification processes. Typically a mid-range to low-range turbidity instrument should be selected to monitor the effluent of a sedimentation or DAF basin. The instrument should at least cover the range of 0.5 to 40 ntu, which is the typical range of turbidity that exits this part of the treatment train.

Particle counting can sometimes be used in place of a turbidimeter at the exit of the sedimentation or DAF basin, if the basin removes the majority of large particles (those greater than 100 µm) and has a turbidity that is no greater than 5 ntu. Particle counting measurements can be used to troubleshoot problems in these treatment units through the profiling of particles that remain after treatment. Particle counters can also be a valuable tool to help evaluate new chemical treatment strategies for both flocculation and sedimentation. Care should be taken, as particle counters are subject to plugging, flow disruptions, and sensing errors when high-turbidity samples are analyzed. Furthermore, some floc particles may be broken up when passing through the sensor, depending on floc size, sensor cross-sectional area, and velocity of the water in the sensing zone.

pH. As discussed above, pH measurements are important for coagulation optimization. pH is less important for clarification process control, although pH probes used for coagulation control are often located in, or downstream of, clarifers. For maximum clarification efficiency, pH should be kept stable through the coagulation, clarification, and filtration processes.

Filtration

Turbidity and particles. Filtration represents the final physical treatment barrier to pathogens in most water treatment plants, and both turbidimeters and particle counters are often used in monitoring filtration performance. Turbidimeters have historically been used to assess filtered water quality, and continuous online monitoring of filter effluent turbidity is required by most regulatory agencies. For plants using chemical pretreatment and rapid-rate granular media filtration, the USEPA requires filtered water turbidity to be monitored at each filter in service at intervals no longer than 15 min. The practical implication of this is that online turbidity measurement must be employed, as collecting and measuring grab samples every 15 min is too labor-intensive to be feasible.

Turbidimeters and particle counters also can be used as filter optimization tools. Assessing filter performance for turbidity and particle removal with these online tools is generally recognized as the best measure of efficiency for the removal of parasites such as *Cryptosporidium* (Bellamy et al. 1993). New low-range, laser-based turbidimeters and particle counters have also been shown to be able to detect particle breakthrough events that were not detectable by conventional turbidimeters. Pilot studies have demonstrated that if used correctly, laser turbidimeters and particle counters can be used to identify coagulant dosages and operational conditions that extend filter run time and reduce filter ripening time.

A case study on use of statistical methods to interpret turbidity data, "The Application of Simplified Process Statistical Variance Techniques to Improve the Analysis of Real-Time Filtration Performance," may be found in chapter 7.

pH. For maximum filtration efficiency, pH should be kept stable through the coagulation, clarification, and filtration processes to prevent precipitated coagulant floc particles from redissolving because of pH changes. This is especially important for controlling aluminum residuals in the finished water. Maintaining pH as close as possible to the pH of minimum solubility for the coagulant used will ensure that the maximum amount of added Al remains in the solid phase where it can be removed by filtration. pH may be adjusted after filtration for corrosion control purposes. Online pH sensors are often installed downstream of filtration to measure finished water pH.

TOC and UV absorbance. As mentioned above, the online monitoring of TOC or UV absorbance can be a powerful tool for coagulation optimization. In the absence of proper coagulation, good filter performance is difficult to achieve, regardless of the filter condition or the use of best filter operating practices (Bellamy et al. 1993). Online UV absorbance or TOC analyzers are often located downstream of the filtration step because of reduced instrument maintenance due to the lower concentration of suspended solids present after filtration.

Differential pressure sensors. Differential pressure sensors are used to measure head loss in filter beds. Their use is most common in rapid-rate granular media filters that employ constant rate filtration with rate-of-flow effluent control valves. Attainment of terminal head loss signals the need to remove a filter from service so it can be backwashed. Operating a filter to excessive head loss can result in turbidity breakthrough or a decrease in flow through the filter when the driving head across the filter is insufficient to maintain the desired production rate in the filter. Knowledge of the rate of increase of head loss can be a guide to operators at plants where filter aid polymers are used, as an excess dosage of filter aid can cause head loss in the filter bed to increase too rapidly by causing removal of strong floc to occur at the top of the filter bed rather than within the bed. Buildup of a mat of floc on the filter bed surface results in cake filtration and a pattern of accelerating head loss with time.

Differential pressure sensors also are needed in contact adsorption clarifiers (roughing filters that use coarse filter media) because deposition of floc in clarifier filters causes head loss to build up and eventually these filters also have to be backwashed.

Plant operators should consult the manufacturer's bulletin for details on maintenance of differential pressure sensors.

Use of multiple sensors. Use of multiple online sensors for monitoring and controlling process performance in water filtration plants is becoming more common. An example of this is presented in the case study, "Online Monitoring Aids Operations at Clackamas River Water," found in chapter 7.

TURBIDIMETERS: TECHNICAL DETAILS

Introduction

Turbidity has long been an important indicator of water quality. In the drinking water industry turbidity is defined by how particles in water interact with light passing through the water. The American Public Health Association's publication of *Standard Methods for the Examination of Water and Wastewater*, twenty-first ed. (APHA et al. 2005), referred to in this manual as *Standard Methods*, defines turbidity as "an expression of the optical property that causes light to be scattered and absorbed rather than transmitted with no change in direction of flux level through the sample." Essentially, turbidity has become used as an analytical measurement for water quality through which light scattered by particulate matter is quantified. In the analytical method used for drinking water, absorption, either by the particulate matter or the sample matrix, is treated as an interference, and many technologies today can reduce or eliminate this interference.

Turbidity in water is caused by the presence of disperse, suspended solids—particles that are not in true solution and often include silt, clay, algae and other microorganisms, organic matter, and other minute particles. Solids in drinking water can support growth of harmful microorganisms and reduce effectiveness of chlorination, UV disinfection, and other disinfectant strategies that can result in compromised health impact to the consumer of such water. In almost all water supplies, higher levels of suspended matter are unacceptable for both aesthetic and health reasons, and they can also interfere with chemical and biological analytical practices.

Turbidity is undesirable in drinking water, plant effluent waters, water for food and beverage processing, and for a large number of other water-dependent manufacturing processes. The removal of turbidity is accomplished through the processes of coagulation or enhanced coagulation, clarification, and filtration. The turbidity measurement provides a rapid means of assessing the turbidity removal process and to assess when, how, and to what extent the water must be treated to meet quality goals or regulations.

Turbidity measurements are the standard tool among drinking water utilities to determine the performance of the particulate removal processes. Studies have shown that the removal of turbidity correlates to a reduction of pathogenic risk associated with human health. In practice, lower turbidity levels translate to reduced *Cryptosporidium* levels and higher water quality (Huck et al. 2000, Emelko et al. 2000). Thus, turbidity is a regulatory parameter that is used to gauge filter effectiveness and to help maintain consistency in the quality of filtered drinking water. Turbidity limits have been in place for decades, and as drinking water regulations have become more stringent, so have the turbidity limits for utilities. The parameter has universal usage for filtration performance.

Utilities under the jurisdiction of the USEPA are required to maintain filtration performance that is monitored through the use of turbidity measurements. For plants using coagulation and granular media filtration, the current turbidity limit is set at 0.3 ntu for 95 percent of the measurements that must be taken with a frequency that is no greater than 15 min (USEPA 1998a). Utilities have demonstrated that the combination of proper operational parameters of their filters, optimized prefiltration techniques (e.g., coagulation and clarification), and well-maintained turbidimeters can achieve these regulatory requirements. In fact, a large number of utilities have internal performance goals to never exceed turbidity levels above 0.1 ntu, and sometimes lower.

Typically, most plants will operate with a combination of online, laboratory, and portable instruments. Each type of application requires different operational features for these turbidimeters. Each type of instrument application is discussed in further detail.

With each type of application for turbidity measurement during the water treatment process, several interferences must be considered, and if present, they must be eliminated or minimized to achieve measurement quality and accuracy. Major interferences in turbidity measurements that are most common in various applications (online, laboratory, or portable) include: dissolved color, particle absorption, stray light, bubbles, sample cell imperfections, and condensation.

History

Early turbidity monitoring.   A review of the history of turbidity measurement can be useful for persons who search older literature to learn about water treatment in the early to middle 1900s. Modern practical attempts to make turbidity a quantifiable measurement date to the early 1900s when Whipple and Jackson (1900) developed a standard suspension fluid using 1,000 ppm of diatomaceous earth in distilled water. Dilution of this reference suspension resulted in a series of standard suspensions used to derive a ppm-silica scale for calibrating contemporary turbidimeters.

Jackson applied the ppm-silica scale to an existing turbidimeter, which was called a diaphanometer, creating what was known as the Jackson candle turbidimeter. This instrument consisted of a special candle and a flat-bottomed glass tube. The tube was calibrated in graduations derived from the ppm-silica suspension. The units were referred to as Jackson turbidity units or JTUs when this device was used (Sadar 1998).

Measurements were made by slowly pouring a turbid sample into the tube until the visual image of the candle flame, viewed from the open top of the tube, diffused to a uniform glow. Visual image extinction occurred when the intensity of the scattered light equaled that of the transmitted light. The depth of the sample in the tube was then read against the ppm-silica scale, and this turbidity was recorded. This method was suitable for raw water samples but was not sufficiently sensitive for filtered water.

Several other devices have been developed for measuring turbidity. Common methods included pouring a solution into a tube and watching for the disappearance of an image that is at the bottom of the tube. This was intended to be an improvement over the Jackson candle turbidimeter in that it eliminated the need for the candle. Several versions of flat disks, called Secchi disks, were also developed. The heavy round disk consisted of four quadrants that were alternated between black and white. The disk was connected to a scaled cable or rope and dropped into a body of water. The point at which the disk disappeared was then correlated to a turbidity value for that body of water. This practice is still in common use for the monitoring of static environmental samples, but it has limited application because of its lack of sensitivity at lower turbidity levels. A variation of the Secchi disk approach at filtration plants was to construct a sight well containing a deep column of water, through which filtered water continuously passed. The bottom of the sight well typically consisted of black and white ceramic tiles. When the tiles could be seen distinctly, filtered water turbidity was low, and as with the Secchi disk, higher turbidity tended to obscure the boundaries between black and white tiles.

Development of nephelometry.   Historically, more precise measurements at very low turbidity levels were needed, as the candle and glass tube turbidimeters were not capable of measuring turbidities below about 4 JTU, which are roughly 16 turbidity units (Sadar 1998). Further, such instruments were highly dependent on human judgment and training to deliver the JTU result. Thus, the visual extinction methods gave way to electronic turbidity methods, which used photoelectric detectors that were very sensitive to changes in light intensity. These instruments provided better precision under certain conditions, but they were still limited in their ability to measure very high or low turbidities. At low turbidity levels, the net change in transmitted light, viewed from a coincident view, was so small that it is virtually undetectable by any means. Typically, the signal was lost in the electronic noise. At higher concentrations, multiple scattering interfered with direct scattering.

The solution to this problem was to measure the incident light scattered at an angle to the incident light beam and then to relate this angle-scattered light back to the sample's actual turbidity. A detection angle of 90° is considered to be very sensitive to light scatter across a wide range of particle sizes. Most modern instruments measure 90° scatter and are called nephelometers if their primary light scatter detector is at this angle relative to the incident light beam.

Theory of Operation

The key criterion for components of a nephelometer is a detector that is positioned geometrically at an angle of 90° relative to the centerline of the incident light beam, as shown in Figure 3-1. Different light sources can be used, which are often dictated by regulatory requirements.

As compared to other methods for turbidity measurement, nephelometers have a significant increase in sensitivity, precision, and applicability over a wide particle size and concentration range. The nephelometer has been adopted by *Standard Methods*, USEPA, and ASTM International as the preferred means for measurement of turbidity. Likewise, the preferred expression for turbidity measured by using an instrument of this design is the nephelometric turbidity unit or ntu (ASTM International 2007). Nephelometers can detect light scattered by particles in the 0.1- to 1-μm size range, with a peak response at about 0.2 μm.

Today, there are several versions of nephelometric turbidimeters, and efforts are under way to distinguish between the different types. The primary differences in design are based upon the type of light sources used and whether or not the method uses a ratio technique. In a ratio turbidimeter, the output of the primary detector is in the numerator of the ratio algorithm, and also possibly in the denominator of a ratio algorithm.

Different nephelometer technologies can produce different results. Efforts have been under way to assign specific traceable units for those nephelometric methods that employ a specific type of technology. The most common units for the measurement of filter effluent are as follows (ASTM International 2007):

- **ntu**: Tungsten filament lamp that is operated at a specific color temperature, and one or more detectors, of which the 90° detector is the primary detector in the ratio measurement. These designs are specified in EPA 180.1 (USEPA 1993) and Standard Method 2130B (APHA et al. 2005).

- **FNU**: This requires the use of an 860-nm incident light source with a bandwidth not to exceed ±30 nm. The technology uses one or more detectors, of which the 90° detector is the primary detector in the ratio measurement. This design is specified by the International Organization for Standardization

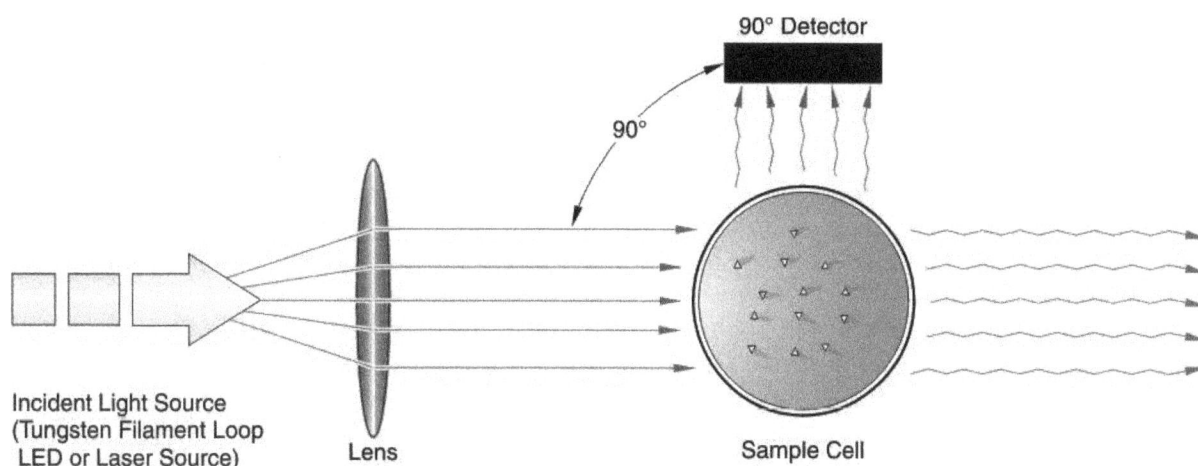

Source: Hach Company, Loveland, Colo.

Figure 3-1   Basic design of a nephelometer

(ISO) Method 7027 (ISO 1999) for turbidity measurement in those regions governed by ISO regulations.

- **mntu**: This unit traces to a laser turbidity method in which a laser diode of 660 nm is used as the incident light source. The technology uses one or more detectors, of which the 90° detector is the primary detector in the ratio measurement. The method is known as Hach Method 10133 and is USEPA approved for reporting in drinking water utilities (USEPA 2002).

The technologies described above represent the current variety of modern instrumentation that is used to monitor filtration performance in drinking water plants. These instruments have been designed to perform with the greatest accuracy and precision at the lowest turbidity levels, which are in the 0.05- to 5-ntu range. Since most regulations require the turbidity levels to be at or below 0.3 ntu (USEPA 1998a) in plants employing coagulation and rapid rate filtration, it is of greatest importance that such instruments provide the stability, low stray light, and excellent sensitivity to the finest changes in turbidity of the water that exits a filter. Note that not all designs are approved globally. For filter effluent monitoring that is used for reporting purposes, stringent design criteria must be met. Ensure that these conditions are met if the turbidity data are used for regulatory reporting.

Current turbidimeter design.   From 2000 to 2010, a significant advancement in turbidity measurement has been observed. Though today's instruments meet the same basic requirements of nephelometers, advanced electronics, ratio algorithms, techniques to eliminate stray light interference, and software improvements have produced modern measurement technologies that are far more accurate and stable. In addition, the new designs may have a greater dynamic range and can be used to measure samples with more complex matrices and be able to minimize the effects of typical interferences.

The most sensitive turbidimeters for applications involving very low turbidity are laser turbidimeters. These instruments use laser-based light sources that project a highly columinated light beam into the sample and create a high energy-density analysis volume. This creates an instrument very sensitive to very small changes in turbidity and high accuracy at low levels. These instruments typically do not have a high dynamic measurement range but have the ability to sense very fine changes in measurement, changes that could be precursors to a major filtration upset.

Such instruments are also common in membrane filtration applications, where a breached membrane element is subject to high levels of dilution. Under such conditions, a highly sensitive turbidity measurement is critical to detection of such a membrane failure (Sadar et al. 2003).

Summary of modern turbidity test methods.   As mentioned previously, the optical property expressed as turbidity is measured by the light scattering effect of suspended constituents within a sample; the higher the quantity of scattered or attenuated light, the higher the turbidity (APHA et al. 2005). Current methods are based upon a comparison of the light scattered or attenuated by the sample with the amount of light scattered or attenuated by a reference suspension under the same environmental conditions. The common components for most turbidimeters include:

1. Light source for illuminating the sample. Typically the characteristics of the light source are specified by regulatory agencies. Light sources can be a polychromatic (incandescent), laser diodes, or narrowband light emitting diodes (LEDs).

2. Sample cell or chamber. The sample cell must allow incident light to pass through the sample and scattered light to the detector.

3. Light detector(s) located to detect light scattered by particulate material. The spectral response of the detector should be matched to the spectral output of the incident light source to generate a usable signal with the desired sensitivity.

4. The instrumentation must have the electronic hardware and software to manage the optical output of the light source, to be able to convert detector response to a turbidity measurement, and to transmit or display the result.

Some designs allow for the sample cell to be external to the instrument and placed in the process flow. These are typically probe instruments and are designed to measure in situ and do not require special sampling protocols to bring sample to the instrument.

For other designs, sample flows into a defined space (sample cell) where the analysis takes place. These designs are referred to as sidestream or slipstream instruments, and because they typically shield the sample from external interferences, they are capable of producing high accuracy and precision at low turbidity values.

Last, bench-top and most portable designs provide a measurement cell, which is manually filled with sample. The cell is placed into the turbidimeter where the measurement is then taken. These instruments are typically designed to operate under known environmental conditions and are capable of producing high accuracy and precision.

Turbidity Standards and Instrument Calibration

To make turbidity a quantifiable measurement, it was necessary to develop a standard that could be reproducibly prepared from defined raw materials. The earliest standards, such as the diatomaceous earth, or fuller's earth, kaolin, and stream-bed sediment were all prepared from materials found in nature, and consistency of such standards was difficult to achieve. Kingsbury et al. (1926) developed a compound known as formazin, which was thought to be an ideal suspension for turbidity standards. The suspension is prepared from accurately measured masses of hydrazine sulfate (5.00 g) and hexamethylenetetramine (50.00 g) that are dissolved sequentially into 1L of water. Under strict environmental conditions of 25°C, the solution develops a white turbidity standard in the course of 24–48 hr. The suspension, when prepared properly and with assayed raw materials, can be produced with a repeatability of 1 percent. Formazin is currently the only true primary standard available for turbidity calibration and measurement, and all other standards and surrogates are traced back to primary formazin standards (Sadar 1999).

Formazin has several desirable characteristics that led to its being the recognized primary standard for quantifiable turbidity measurement. First and most important, it can be reproducibly prepared from assayed raw materials. Second, the physical characteristics of the light-scattering polymer produce consistent light scatter with little to no absorbance. Third, the polymer consists of chains of different lengths, which fold into random configurations, producing a wide array of particle shapes and sizes in the range of 0.1 to 10 μm. Studies of particle distributions in formazin suspensions indicate irregular distributions among the different lots, but because of the broad distribution of particle sizes and shapes, the overall light scatter is consistent across lots. This randomness of particle shapes and sizes within the formazin suspension yields statistically reproducible light-scattering characteristics for all makes and models of turbidimeters. Thus, most turbidimeters in use today have calibration algorithms that are derived from data that was generated using this standard (Sadar 1999).

When used for regulatory purposes, the definitions of turbidity standards as calibration materials can become confusing, as the definitions used by organizations such as the USEPA (USEPA 1993), ISO 7027 (ISO 1999), and APHA, WEF, and AWWA

(APHA et al. 2005) differ in use and meaning. For regulatory applications under the governance of the USEPA, the definition of a primary standard indicates that the turbidity calibration standard can be used to perform instrument calibrations. This definition has nothing to do with the traditional chemistry definition of a primary standard. The twenty-first edition of *Standard Methods for the Examination of Water and Wastewater* clarified these differences between the regulatory definition and chemical definition of the primary standard, and this manual reflects those clarifications. Currently, there are four recognized calibration standards that can be used for reporting to USEPA for turbidity: Formazin, StablCal® stabilized formazin (Hach Company 1996, USEPA 1997), AEPA-1 AMCO Clear® Styrenedivinylbenzene (USEPA 1997), and SDVB® standards. At the time of this writing, other standards can be used only for verification of instrument calibrations.

Calibration verification standards are those standards that are typically supplied by the instrument manufacturers for checking calibration stability on specific types and models of instruments. Some methods define these standards as secondary standards (APHA et al. 2005). Examples of calibration verification standards include sealed sample cells that contain a light scattering material with an indication of the length of stability, opto-mechanical light-scattering devices, latex suspensions, and metal oxide particles that are trapped inside a gel matrix, to name a few. Under the USEPA definition, secondary standards, once their values are defined, are used to verify the calibration of the turbidimeter for which they were produced (USEPA 1999b). However, these standards are not to be used for calibration itself.

In older literature, procedures were provided that allow for an operator to calibrate an online turbidimeter through the use of a laboratory turbidimeter (APHA et al. 2005). In reality, it is a very difficult task to knowingly gain agreement between the two instruments because of instrument design differences and interferences that are prevalent in low-turbidity samples. Most laboratory turbidimeters will have a higher level of stray light, and this can contribute positive error to exceptionally clean water samples, especially for those in the range of 0.05 ntu and lower. In addition, it is exceedingly difficult to collect and prepare an ultra-clean sample, perform the measurement, and then adjust the online turbidimeter. Typically, this will generate more error than if calibration is performed with defined turbidity standards, with each instrument being calibrated according to the prescriptive procedures that are provided in the respective instrument manuals. In short, calibration of online turbidimeters through comparison is not recommended.

Turbidimeters in the Plant

There are two basic designs of online turbidimeters: sidestream and in-situ. Sidestream turbidimeters are designed to receive a sample into a turbidimeter body or cell where the measurement is performed. The sample then exits the instrument. Sidestream turbidimeters are complete flow-thru instruments in which the sample continuously passes through the instrument. The second designs are in-situ instruments, in which the sensor portion of the instrument is placed directly into the process itself. In-situ instruments do not require sample to be transported to and from the instrument. These two designs have advantages and disadvantages, with some designs performing better in certain applications and worse in others.

Sidestream turbidimeters. Sidestream turbidimeters (sometimes referred to as process turbidimeters) are designed to continuously measure turbidity as a sample passes its optical elements (i.e., light source and detector). The instruments are typically the workhorses for ensuring the processes are in control. For filtration effluent, online turbidimeters can continually show that a filtration system is or is not operating within all regulatory compliance limits, and if a filter is out of compliance, these instruments are typically the first line of defense in notification of the plant staff that a breach in the filtration process has occurred.

Sidestream turbidimeters typically are better for low-turbidity measurements, and most low-range turbidimeters are of this design. The passage of samples into the instrument helps to condition the sample by removing bubble interference as well as preventing the penetration of stray light from outside sources. While these instruments possess the advantages noted above, they are sensitive to flow, pressure, settling of particles, and fouling of optical surfaces. To prevent measurement error due to these variables, the following is suggested:

- Operate the instrument within the manufacturer's recommended flow range. These flow ranges are designed to optimize the removal of bubble interferences and prevent particle fallout. If samples possess a high degree of dissolved gases, a sample flow at the lower end of the specified range is recommended. If particle fallout is of concern, operate at the higher end of the flow range.

- Operate the instrument at the specified pressure range. Some online instruments utilize a pressurized measurement chamber to prevent outgassing of samples and the interference of bubbles.

- The potential fouling of internal surfaces should always be considered. The instrument optical surfaces, namely the detectors and any windows through which light passes, should be inspected to ensure they are free of any deposits, such as oxides of manganese, that could form over time. Clean these surfaces as instructed by the instrument manufacturer. Deviation from these recommendations could lead to damage to these surfaces.

- Inspect all sample chambers or flow-thru cells for particle settling. If settling persists, increase the flow rate and recheck to determine if the problem has been resolved. If increased flows do not solve the problem, another instrument design should be considered such that settling does not occur.

- Instrument settings. Most instruments have a recommended measurement strategy for minimizing noise so that representative data can be consistently generated. These settings include bubble rejection algorithms, signal averaging, and data logging frequency. Refer to the instrument manufacturers and regulatory authorities for recommended or required protocols for setting up these instruments and the logging of measurement data.

- Calibration and verification. Like laboratory turbidimeters, online instruments perform best when specific calibration points are used. The operator should follow the manufacturer's instructions for calibration and if they deviate from these instructions, they should consult with the manufacturer to determine if performance will be compromised. Many instruments also have specific procedures for verification of instrument performance. Be sure to understand how to properly use the verification methods and how to identify if an instrument generates a pass or fail verification. Turbidimeters used to collect data that are required for regulatory compliance must be calibrated at intervals that are no longer than those required by instrument manufacturers or the regulatory authority.

In-situ turbidimeters. These turbidimeters are designed to be placed directly into the process stream and monitor the turbidity of the water sample as it passes by the instrument's sensing elements. These instruments are commonly placed into the open channels or tanks or are mounted in-line or inside pipes. In-situ turbidimeters typically have the light source and the light detector(s) on the same surface but are geometrically positioned (sometimes with the aid of additional optics) to form the required nephelometric angle. This optical geometry provides a wide dynamic-range of measurement but limits the sensitivity of the measurement. Thus, in-situ turbidimeters are typically used for higher-turbidity samples, such as raw waters, filter backwash or other waste streams, and sample points upstream of the filters. These instruments should not be used for monitoring of very clean waters, such as those with turbidities below 0.5 ntu.

Users of in-situ turbidimeters should be aware of several interferences. In addition to the same interferences that affect sidestream turbidimeters, in-situ turbidimeters also have potential interferences from ambient light and from reflections within the sample process. To minimize such interferences, the instrument should be placed away from direct sunlight or in a shrouded environment. These instruments should be mounted in a position so that the incident light beam cannot be reflected off a surface and back in the direction of the sensor, as this will cause significant false-positive error. If mounted within a pipe, the in-situ instrument should face a black surface (or be mounted in a black section of pipe) to minimize reflections. In-situ instruments typically do not have mechanical means of bubble removal, other than being in a pressurized line. If the instrument is not in a pressurized environment, it should be installed at a level where bubble formation is minimal (such as in a deep location), and instrument software algorithms should be used to minimize the effects of bubble interference. Last, when mounting an instrument in a process, the instrument should face the flow so that fouling of the optical surfaces is minimized. Consult the specific manufacturer for best practices when using these instruments.

Comparison between sidestream and in-situ turbidimeters.   The installation to the left in Figure 3-2 illustrates an in-situ application. The face of the instrument should point downstream to prevent scratching of optical components on the face of the instrument. Second, the optical components must be installed at an orientation so ambient light such as reflections or sunlight does not impose an interference in the measurement. The installation to the right in Figure 3-2 is a low-range online turbidimeter. Here, the sample is transported via a sample line to the instrument. The sample flows through the instrument, and the turbidity is measured inside the instrument. After measurement, the sample is sent to drain or back to the process. Sidestream

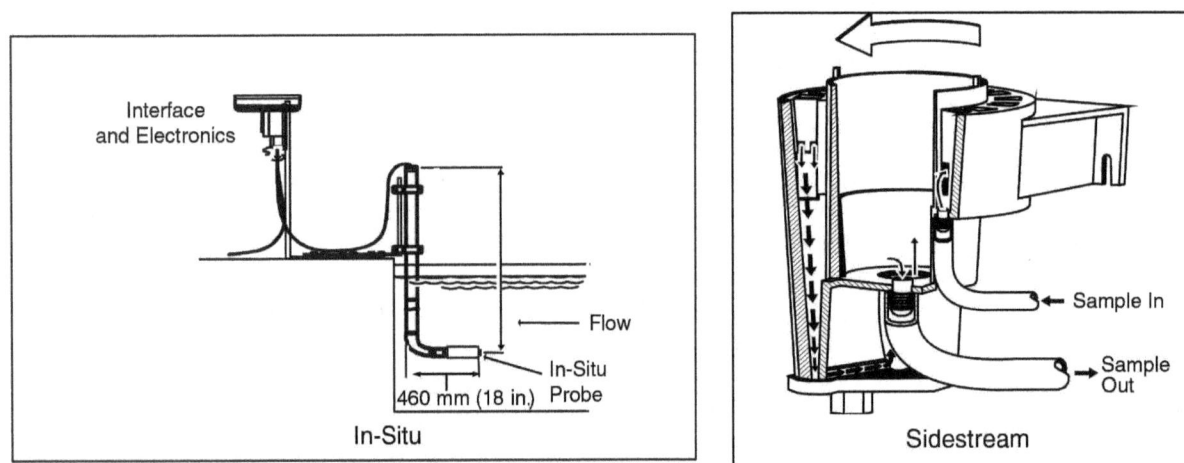

Source: Hach Company, Loveland, Colo.

Figure 3-2   Diagrams illustrating the difference between in-situ and sidestream turbidimeters

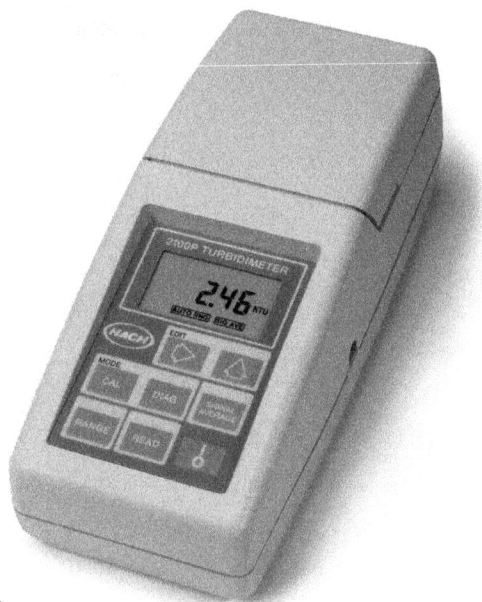

Source: Hach Company, Loveland, Colo.

Figure 3-3   Typical portable turbidimeter

turbidimeters (to the right in Figure 3-2) are designed to eliminate interferences such as light reflections, bubbles, and changes in flow.

Portable Turbidimeters

Portable turbidimeter designs can be highly varied, but many use standard nephelometric and ratio designs. The instruments are usually small, battery-powered and can be taken anywhere to perform measurements, such as the turbidimeter depicted in Figure 3-3.

Depending on the instrument design, portable turbidimeters may or may not meet regulatory parameter protocol. Many designs do not meet the EPA 180.1 design criteria for the light source, so be sure that an appropriate design is being used if reporting turbidity data for compliance with regulations. A host of designs do meet the ISO 7027 criteria for turbidity monitoring.

Sample collection, preparation, and measurement practices for portable turbidimeters are similar to those discussed below for laboratory turbidimeters. Since portable instruments are handled more frequently, it may be worthwhile to increase the frequency of verification checks to ensure performance between calibrations.

Laboratory Turbidimeters

Laboratory turbidimeters, such as the one shown in Figure 3-4, are typically used to ensure the performance of all online turbidimeters in the field. Although online turbidimeters should not be calibrated based on laboratory turbidimeter readings, if the effluent turbidity from one filter in a bank of filters is substantially higher than the turbidity of water from the other filters, a comparison of online and laboratory turbidimeter data for the filters can reveal whether or not both sets of data are similar. If laboratory turbidimeter data show low turbidity for all filters including the one with the high online reading, the high online reading should be considered suspect. In this case, calibrate the turbidimeter giving the high turbidity output, using an approved calibration procedure.

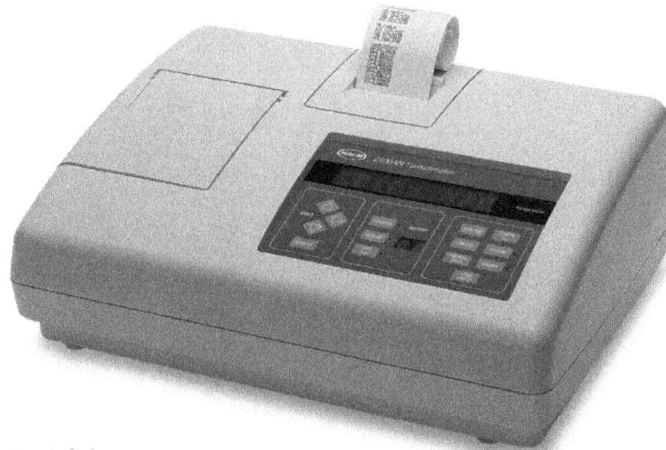

Source: Hach Company, Loveland, Colo.

Figure 3-4   Laboratory turbidimeter commonly used in drinking water plants

Most laboratory analysts view these instruments as being more stable and accurate over time because the instruments are operated and maintained in a controlled environment. However, depending on the design of the online instrument, accuracy and stability may actually be better with the online instrumentation, since the online instruments may have optical designs that minimize the effects of stray light and environmental factors on instrument stability. Instruments that have designs in which the sensors are submerged in the sample tend to exhibit these characteristics.

For most laboratory turbidimeters, samples are manually obtained from the process flow, transferred into a measuring cell, and prepared for measurement (capping, polishing, and degassing). When measuring samples, it is important to ensure that the proper sample preparation procedures are followed to obtain an accurate measurement. Manufacturers of laboratory turbidimeters typically provide detailed information on the preparation of sample cells for measurement. A few of the important facets of the measurement of samples in laboratory turbidimeters are discussed below.

Cleanliness of sample cells.   Ultra clean cells are essential when monitoring the effectiveness of filters. Make sure the sample cells are clean and free from debris—clean as per manufacturer's requirements. For glass cells, washing cells with a soft brush and a combination of water and laboratory detergent should be performed. Follow with an acid wash with a 10 percent HCl solution. Follow by rinsing the cells at least 10 times with water that has been filtered through at least a 0.45-μm filter (or smaller). Cap the cells. Wipe the outsides clean with a soft cloth. Clean cells will exhibit an internal surface that does not allow the formation of water droplets. The formation of water droplets indicates that the surfaces are contaminated with sites that allow water droplets to attach. Some cells may never show this characteristic, and in this case discard the cell and seek an appropriate replacement.

The surfaces of sample cells can cause the reflection and scattering of light through scratches or other imperfections. In such cases, these imperfections can be minimized through specialized cleaning procedures that instrument manufacturers typically provide. In the absence of such procedures, the use of silicone oil will minimize the effects of cell imperfections such as scratches. To polish a cell, apply a thin bead of silicone to the cell. Using a dry, soft cloth, spread the oil over the entire surface of the cell. Follow by wiping the excess oil off, to where it appears the cell is free from oil. At this point the cell is polished.

One of the most common causes of false positive measurements is entrained air in the samples. If the sample contains high amounts of dissolved air, the air can be removed through the application of a small vacuum on the cell or by allowing the cell to stand for several minutes. Then, immediately prior to inserting the sample into the laboratory turbidimeter, gently invert the cell one to two times to resuspend any material that may have settled. This will ensure that the sample is homogenous and provides for an accurate measurement.

Condensation on the outside of the sample cell can cause interference in humid environments when the samples are colder than the surrounding environment. If the cells become fogged because of the accumulation of moisture on the outside of the sample cells, they must be wiped until the sample warms and the condensation stops. Though it is traditional practice to measure the sample at the temperature at which it was captured, the error in the sample due to a temperature change will be far less than an error due to condensation. With some instruments, dry instrument air can be blown across the outside of the sample cell to prevent condensation.

Color and absorbance can also result in negative interference in turbidity measurement. This interference cannot be eliminated through sample preparation procedures, but it can be minimized through the application of specialized turbidimeter measurement technologies. Color and absorbance can be minimized by two different measurement protocols. The first is to use a wavelength of incident light that is not absorbed by the constituents within the sample. Infared (IR) light sources have been found to be the most effective at minimizing these interferences but do not eliminate interference in all samples. Thus, the analyst must perform spectrophotometric checks on the samples to ensure that the light absorbance within the sample is at a wavelength different from that of the incident light from the turbidimeter.

A second method of minimizing the impacts of color and particle absorbance is through a ratio technique. A ratio technique involves the use of a second light scatter detector that measures the amount of light that is attenuated away from the nephelometric detector, as shown in Figure 3-5. A software algorithm then calculates the amount of light lost due to absorbance and corrects the result. Ratio techniques have been available for approximately 20 years, and the regulatory authorities typically approve the methods if the primary detector is a nephelometric detector. Most modern laboratory turbidimeters have a ratio feature. Two or more detectors may be present in

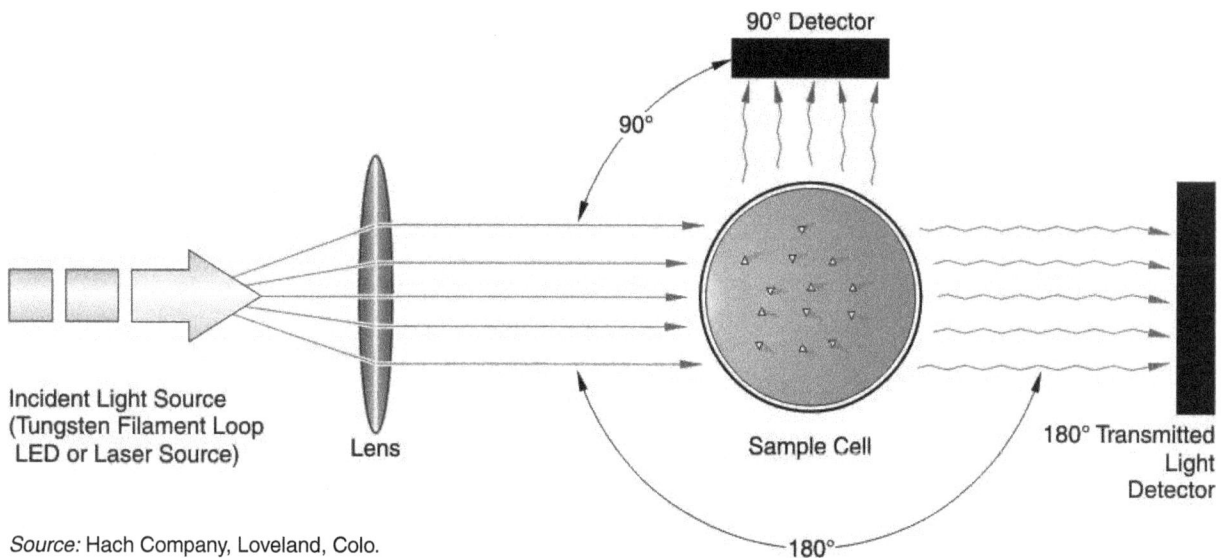

Source: Hach Company, Loveland, Colo.

Figure 3-5   Optical geometry for a basic ratio system involving two detectors

different designs to help reduce various interferences, adjust for incident light variation, or extend the measurement range of the instrument.

Optimization of laboratory turbidimeters. Laboratory turbidimeters are, by default, assumed to be the correct reference instrument against which other instruments are benchmarked. They are commonly used to verify performance of other turbidimeters, such as those that continuously monitor processes within the different stages of the drinking water treatment train. Many plants, especially those that are of smaller size, have used laboratory turbidimeters for regulatory reporting. In the present regulatory environment, if filter effluent turbidity must be monitored at intervals not exceeding 15 min for each filter, this use of laboratory turbidimeters is not practical from the standpoint of labor required for the analyses. Laboratory turbidimeters are capable of covering a broad operating range and also incorporate compensation features for the reduction of interferences such as color and bubbles. A large majority of plants also have laboratory turbidimeters integrated into their standard operating procedures for verifying calibration of other turbidimeters (online and portable), and for checking the accuracy of standards that are used to either calibrate or verify measurement performance. In short, the laboratory turbidimeters are the key reference turbidimeter for most plants and weigh heavily on ensuring the performance of the treatment processes is maintained. Thus, it is critical that laboratory turbidimeters be set up and maintained in optimal operating condition.

Laboratory turbidimeters should be treated like other laboratory instruments, in that they should be set up in a clean and stable environment. The location of setup and operation should be free of dust and debris, away from direct sunlight or sources of heat, and in an area where the temperature is stable. The location will ensure that the instruments maintain stability and accuracy at the critical low end of the measurement range—the area where regulatory compliance is a critical goal of the water treatment plant.

Laboratory turbidimeters should be calibrated accordingly to the manufacturer's instructions, which for most instruments include specific calibration points and calibration standards (USEPA 1999b). When manufacturers design turbidimeters, the calibration algorithms are typically built around the use of standards with specific values that will deliver the most accurate measurements over the range of interest. Deviation from these manufacturer-recommended procedures increases measurement error of these instruments. Upon the completion of calibration, verification of calibration should be performed, using guidance from the manufacturers. Most modern instruments employ features to increase measurement stability over time, but a sudden contamination of critical measurement components or an accident involving the instrument could change the calibration. Thus, verification is a key asset to ensuring accuracy of measurements. Verification procedures are typically suggested by manufacturers and may involve the use of wet standards, dry standards, or contained standards. Regulations, which can be regional, state, or local, typically do not specify the type of verification standards, but they do typically specify the frequency of verification, which ranges from one week to one month.

Standard operating procedures should be developed to ensure that maintenance is conducted at timely intervals. Maintenance includes cleaning of sample compartments, cleaning of the area surrounding the laboratory turbidimeter, lamp replacement (based upon verification), and the cleaning of ancillary equipment that is associated with the laboratory turbidimeter. This equipment includes sample cells, sample collection flasks, and laboratory equipment that is used to prepare calibration and verification standards.

The combination of optimized operating conditions, accurate calibrations that follow manufacturers instructions, verification, and maintenance standard operating procedures (SOPs) will provide a laboratory turbidimeter that can be reliable and serve as an excellent reference instrument against which other processes and instrumentation can be monitored.

Best Practices for Sidestream Turbidity Analysis

Successful turbidity measurement is achieved with the combination of proper instrument selection, routine calibration and verification, and the use of best measurement practices. Some general guidelines can be followed to ensure that the turbidity measurements are representative of the sample and that these measurements provide the operator with information to be proactive regarding their processes.

- Instrument selection: Select the best technology to perform the job based on the turbidity range of the process and the potential interferences in the sample. For example, do not select a laser turbidimeter that is designed for low-level measurement if the sample is upstream of the filtration process. If the instrument is used for regulatory compliance, ensure its design meets the specified criteria that are outlined by the regulatory compliance method. Do not assume the one-size-fits-all approach when selecting an instrument.

- Location: Ensure that the instrument is located as close to the sample as possible but also at a location where proper calibration and maintenance can take place. It makes no sense to place an instrument in an inaccessible location to minimize the sample lines if this instrument cannot be easily calibrated and maintained. Given the choice between short sample lines and accessibility to the instrument, accessibility takes precedence.

- Sample line for filter effluent: Place the sample line in the middle of the effluent pipe to avoid air bubbles at the top of the pipe and sediment at the bottom, upstream of pipe bends and valves, and upstream of the diversion point for filter-to-waste, if this feature has been provided. Sample tubing should be noncorrosive and should not support biofilm growth, as sloughing of corrosion products or biofilm could cause high turbidity readings not indicative of filtered water quality.

- Perform calibrations as instructed by the manufacturer. The instructions were developed by persons having thorough knowledge of their instruments and how to optimize instrument performance. The operational procedures that are provided by the respective manufacturers should be followed.

- Prior to selection of the instrumentation, make sure that the data collection and transmission method is appropriate for the utility's supervisory control and data acquisition (SCADA) system.

- Understand the requirements for calibration verification and maintenance. Understand the time and cost requirements to perform all maintenance and verifications that are necessary to comply with regulations and to ensure the monitoring of the process is adequate.

- When selecting a laboratory turbidimeter as a reference, it is imperative that performance between the two instruments be quantified. Many online designs will read slightly lower and more accurately than the laboratory reference at extremely low turbidity levels (<0.1 ntu). Understand what these differences are when both systems are optimized. Once this bias has been determined and applied, comparability between online and laboratory measurements will be more consistent and improved.

- Understand the environment under which the online instruments are to be operated. Environments that have high temperature extremes, high humidity, and corrosive environments should be identified and appropriate instrument designs that are tolerant to these conditions should be selected.

- Maintenance: All instruments require maintenance and turbidimeters are no exception. At the very least, instrument sample chambers should be flushed (according to manufacturer's instructions) at least monthly or as experience dictates. The higher the turbidity of the sample, the more frequent the flushing cycle. When cleaning instruments, also clean or replace all sample lines, valves, bubble removal devices and flow control devices, and any surface over which the sample flows. Instrument lamps should also be changed when verifications show a decrease in lamp output. As a lamp ages, it will exhibit spectral shifts and output changes. If an instrument continuously fails verification, the lamp should be replaced. Turbidimeters should be calibrated after cleaning and after lamp replacement.

Troubleshooting of Turbidimeters

As with all instrumentation, failures of turbidimeters will take place. However, with a robust program for maintenance and verification, identification of problems or failures will be prompt and remedies can be applied thereafter. Some of the most common turbidimeter problems and how to best troubleshoot them are listed in Table 3-2.

Summary

The measurement of turbidity is commonly used at numerous locations throughout the coagulation and filtration processes of water purification. Common and useful locations include the beginning and the end of clarification and for regulatory applications at the end of each filter and on combined filter effluent streams. Some turbidimeters can also have uses in the specialized filtration mechanisms such as low-pressure membrane filtration.

It is important to understand that there is not a one-size-fits-all technology for online turbidity monitoring. Sample composition and process requirements dictate the type of technology to be used. Upstream in a treatment process (toward the raw water monitoring point) will require higher sample throughput to prevent sample settling and may require long-wavelength or ratio measurement designs to compensate for dissolved colors and particulate absorbance. As the treatment process progresses downstream and the samples begin to clarify, instruments that still have high throughput but more accuracy in the lower turbidity ranges are appropriate. For filtration integrity monitoring and combined filter effluent monitoring, instruments that both comply with regulatory monitoring requirements and have the highest accuracy at low turbidity levels should be used. Last, those processes that perform postfiltration particulate removal processes, such as membrane filtration, may require the most sensitive technologies, such as laser nephelometry or particle counting methods. Applications for turbidimeter technologies are presented in Table 3-3.

Table 3-2   Common turbidimeter problems and troubleshooting approaches

Symptom	Remedy
Online turbidimeter reads a verification standard higher than expected	Clean the turbidimeter to ensure all settled particles have been removed.
Particles continually settle in the turbidimeter	Increase sample flow rates.
Sample has a high degree of noise	Reduce the flow, or increase the pressure of the sample chamber (if applicable), as bubbles are likely not being removed.
The instrument will not verify after calibration	Recalibrate: make sure standards are prepared accurately, make sure the instrument is properly cleaned and flushed, and make sure the calibration instructions are being followed.
A new turbidimeter does not read exactly the same as the old instrument	May be a result of advances in technology or different optical geometry. Consult the manufacturer for the explanation.
Readings continually decrease over time	The lamp may need replacement. Check the optical components for fouling or deterioration. If identified, clean or replace the component as necessary.
Calibration standard in the turbidimeter will not become stable	If the standard was poured into the measurement chamber, it is likely that the chamber was dirty and particulates are causing the poor stability. Flush everything and recalibrate.
Significant bubbles form on the internal surfaces of the instrument chamber	The surfaces are not becoming wetted. Soak the surfaces with an oil-cutting detergent (dish soap or laboratory detergent) for several hours to several days.
Values on the laboratory turbidimeter read significantly higher than the online instrument	Check to make sure condensation is not forming on the turbidimeter sample cells. Determine if the sample has color. If so, the online turbidimeter may not be compensating for color and the laboratory turbidimeter is providing a correction. Process turbidimeters that eliminate the use of glass sample cells eliminate light scatter off these surfaces and produce a lower, more accurate reading than laboratory turbidimeters. In general, laboratory turbidimeters should read within 0.03 ntu of most process turbidimeters that are correctly calibrated.
Online turbidimeter reads higher then laboratory or portable turbidimeter	Check the calibration of both instruments. For instruments for which a zero standard is measured, there is a tendency to overcompensate for this value, which pulls the values of the instrument falsely negative. Check instruments with a defined verification standard in the 0.1- to 0.5-ntu range to determine which instrument is right and which instrument is in error. Liquid verification standards are commercially available.

Courtesy of Mike Sadar.

Table 3-3   Appropriate application for given turbidimeter technologies

Location in a Drinking Water Plant	Technology Considerations for Successful Monitoring
Pre- and Postclarification	Instruments must have a wider measurement range that can cover the turbidity from effective particulate settling, which is typically down to 0.5 ntu. These instruments require a high level of sample throughput to prevent settling inside the instruments, or probe-style instruments that are installed in the correct location and are not impacted by ambient light are often preferred.
Backwash	An instrument should have a range of at least 0–1,000 ntu and be of a probe design that is capable of providing a rapid response to a backwash as spent backwash water begins to clarify. The location of the instrument is critical to capture the correct turbidity for backwash termination. The system should also be immune to ambient light or reflection and should utilize near infrared light (800–900 nm) so it maintains sensitivity in the presence of highly absorbent particulates.
Individual Conventional Filtration and Combined Effluent Monitoring	Turbidimeters that have focused performance for accuracy in the range of 0–5 ntu and meet regulatory design requirements. Instruments should be able to compensate for bubble interferences and not be subject to particulate settling. Instruments should have effective verification accessories for regulatory compliance.
High-Performance Filtration (e.g., Membranes)	The most sensitive turbidimeters with high-end accuracy at low levels are needed. Instrument typically must meet regulatory design requirements for reporting and be able to eliminate bubble interferences. Laser turbidimeters and particle counters are common technologies.

Courtesy of Mike Sadar.

PARTICLE COUNTERS: TECHNICAL DETAILS

Particle counters have been used in water treatment research and pilot plant filtration studies for about four decades. For drinking water studies and plant monitoring, use of instruments that can count particles and measure their size using the principle of light obscuration began in the 1970s. Many types of particle counter technologies exist, but light obscuration particle counters are the type most commonly used to monitor the effectiveness of sedimentation and filtration performance, and this technology will be discussed in this chapter. Light obscuration technologies are capable of sizing particles down to about 2 μm, and a few designs can size down to a minimum of 1 μm. Particle counters that are capable of sizing particles in the submicron level do exist but are not used in drinking water for optimization because of excessive cost of ownership, maintenance, and operation.

Theory of Operation

Light obscuration particle counters (see Figure 3-6) are based on the ability of a particle to obscure or scatter light as it passes between a optimally controlled light source and a detector. In this design, a flow stream passes between a highly columinated light source (typically a laser diode that emits monochromatic light) and its detector, which are at a defined distance apart inside a geometrically defined area called the flow cell. As a particle passes between the light source and the detector in the flow cell, it both absorbs and/or scatters a quantifiable amount of light away from the detector. This results in a net decrease in the amount of light that would otherwise reach the detector if the particle was not present. Theoretically, a greater amount of light that is either scattered or absorbed by a passing particle correlates to a greater decrease in the detector current. And a larger decrease in detector current correlates to a larger particle. The number of particles that pass over the detector in a given volume of fluid

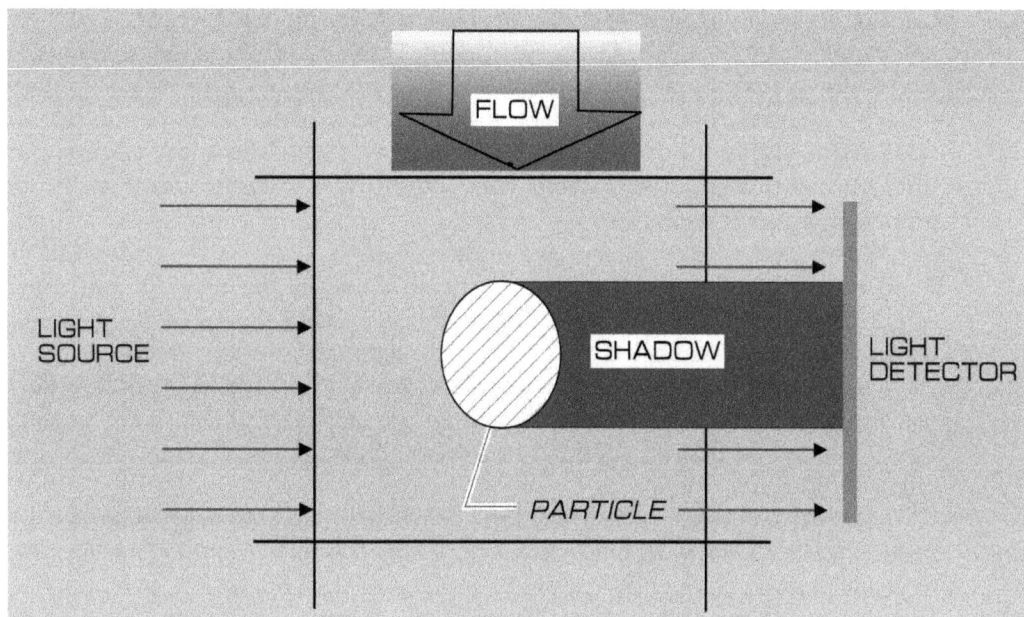

Source: Hach Company. Loveland, Colo.

Figure 3-6   Theory of operation for a light obscuration particle counter

(detector events) also correlates directly to the number of particles present at that particular time. Thus, to a given limit of concentration, light obscuration particle counters can provide both particle size and particle count.

Counting and sizing principles of light obscuration particle counters. Particle sizing is done in reference to calibration suspensions, which typically are composed of round polystyrene latex (PSL) particles having a given refractive index. Real-world particles differ in their respective shape, refractive indices, and absorptive characteristics. Thus, when samples are analyzed by a particle counter, the sizes are not absolute but are relative to the calibration materials used to develop the particle sizing calibration curves in particle counters. Particle counters also lose size sensitivity in the 1- to 2-μm size range, with the lower limit being dependent on the design of the flow cell. Light obscuration particle counters are not capable of accurately sizing or counting particles that fall below the lower limit of detection.

Particle count levels are valid only for a given range of particle concentration. As the particle concentrations increase, they will eventually reach a point where two separate particles cannot be distinguished from each other as they pass through the flow cell. This is often referred to as particle overconcentration or coincidence. The level of coincidence will differ among samples and is dependent on the particle size distribution of a given sample. The coincidence limits are typically approached in the 5,000 to 15,000 particles per mL (cts/mL) range. This value of coincidence can vary among manufacturers and flow cell designs.

It is important to understand both the count and size limitations of a particle counter. For most applications in drinking water, particle counters should not be used if the count concentration range exceeds the 50 percent point of the manufacturer's stated coincidence limit. For example, if the coincidence limit specification is 10,000 cts/mL, then the practical use limit should be for samples with a particle count limit that does not exceed 5,000 cts/mL. With respect to the limit of sizing, it is important to understand the accuracy of a particle counter (expressed as resolution) and the cutoff point for the smallest particle size. For best reliability with respect to size, a limit of detection should be determined, which can be on the order of 1.5 to 2 times the stated size limitation of the instrument. If the size limit is too high for the respective water plant's application, then a lower limit can be determined if a verification procedure exists that can prove both count and size accuracy at the bottom end of the particle sizing range for a given instrument. Particle counter manufacturers will have verification procedures that can be used to determine these levels of performance for a particle counter.

The accuracy of particle sizing and counting is also very dependent on the ability to accurately monitor the volume of fluid that passes through the instrument's flow cell. Particle counts are determined per unit volume of fluid that is determined when the instrument is calibrated. If the flow rate through a sensor is inaccurate, so will be the particle counts. Most online instruments are set up with constant flow devices that operate on a principle of discharge through an orifice under constant head to ensure that a continuous flow at a defined rate passes through the flow cell at all times, which ensures count accuracy.

Keys to the correct application.   Particle counters are very sensitive water monitoring tools that can yield information regarding the performance of a treatment process in sedimentation through filtration. Regardless of the application, several key sampling and monitoring requirements are necessary for optimized use of the technologies. These include: sample type (appropriate versus inappropriate), sampling points, sample lines, flow rates, cleaning, interferences, performance verification, and calibration.

Appropriate samples.   The most appropriate samples to be monitored by particle counters are at the filter effluent and downstream from the final filtration process. At this point of the treatment process, the particulate level particle concentration can be used to monitor the performance of the filters and to monitor for a particle breakthrough in filtration or deterioration of the finished water quality as it progresses into distribution. Particle concentrations in these samples are typically well beneath the coincidence limits of a given technology and so are less prone to plugging the flow cell. When used in these applications, maintenance programs are defined and proven to be effective.

Prior to filtration, such as at filter influent, particle counting can also be effective as long as the particle concentrations in the samples are beneath the coincidence levels of the instrument and a screening filtration system is part of sample pretreatment. Without the prescreening, the flow cells are more prone to maintenance needs and can plug. Depending on floc size and strength and flow cell cross section, large floc particles may be broken up during passage through the flow cell, creating spurious data. Sample matrices should also be free from colloidal materials and high levels of color, both of which can yield erroneous results. Biologically active samples can be monitored but will require more frequent cleaning and maintenance.

Inappropriate samples.   Samples that are high in particle content, such as those that exceed a turbidity of about 3 to 5 ntu, will typically be too concentrated for use with light obscuration counters. Higher-turbidity samples can often exceed coincidence limits and give false-positive sizing information (particles undergoing coincidence can be counted as a single bigger particle). Further, higher-turbidity samples typically will have larger particles that can plug the flow cells of some technologies. Flow cells are typically on the order of 0.50- to 1.0-mm diameter maximum, and if particles approach the diameter of a flow cell, the cell can easily become plugged. Samples that contain entrained air can also be problematic. A prescreening filter can be used to reduce the probability of plugging a flow cell.

Air bubbles also scatter light and will generate false positive results. This is often the case when monitoring membrane systems that use air scour processes to reduce fouling levels. Backwashes can also introduce entrained air into sample lines, which will generate false positive values. These interferences are caused by the injected air and can be reduced or even eliminated if adequate bubble removal devices are employed and the timing of backwash events can be synchronized so that monitoring does not take place during and for a defined period of time after a cleaning event.

Sample points.   Sampling is key to obtaining results that are directly representative of the treatment process. Sampling should be at a point where the sample is completely homogeneous such as the middle of the filter effluent piping, upstream of valves and bends. Sample points should be away from any chemical or air injection point so that bubble interference is minimized. Last, sample points should be in close proximity to the instrument flow cell to improve response time and to shorten the duration of any known bubble event (air scour or backwash). Sample points should also be at a location that enough flow is provided to a particle counter without the use of a pump. Pumps should never be used to push a sample from its source and the turbidimeter. If a pump is needed, it should be located on the downstream side of the sensor and it should draw the sample through the particle counter flow cell first so that pump interferences are minimized (Hargesheimer and Lewis 1995).

Sample lines.   Sample tubing should be composed of inert materials so that particulate sticking and particulate shedding are minimized. Ideally, the best material for sample lines is polytetrafluoroethylene (PTFE) (Hargesheimer and Lewis 1995). If PTFE sample tubing is not available, then select the most inert material that is available. A commonly available example is high-density polyethylene. Also, tubing that is opaque (black in color is best) will keep organisms that require light from growing on the inside surfaces of sample lines. Avoid use of low-density or soft plastics.

Keep sample lines as short as possible between the sample tap and the instrument. Last, sample lines should be replaced according to manufacturer's recommendations, if instrument verification results begin to bias high, or microbial growth becomes apparent. A conservative approach would be to replace sample lines on a semiannual basis.

Flow rates.   Most manufacturers have a defined flow rate at which the particle counter is calibrated. In most cases, this range will be between 50 and 200 mL/min. If possible, try to meet the same flow rate at which the sensors were calibrated, to attain the highest level of size and count accuracy. If the sample flow rate must be changed, the resultant flow rate must also be changed in the particle counter software. Failure to do so will result in significant measurement error. Never use a sample flow rate that is outside the stated limits of the respective technology.

Cleaning.   Particle counters are very sensitive instruments and require adequate maintenance protocols to ensure consistent and accurate performance. Dissolved solids and microbials in samples can lead to the fouling of the flow cell, sample tubing, bubble traps, and flow control devices. As surfaces become dirty, the probability of particle shedding increases and can cause false-positive particle spikes and count levels. The plating of unwanted materials on the surfaces of the flow cells can also cause sizing errors.

Follow the manufacturer's instructions for cleaning of flow cells and use only the apparatus and chemicals that are recommended. Some particle counter technologies have specialized flow cell construction and can be cleaned only by using certain procedures and chemicals. Failure to follow the instrument manufacturer's instructions could damage the instrument. It is often more prudent and economical to replace rather than clean sample lines, depending on the type of cleaning necessary and the time requirements to complete this task. Also, cleaning of the flow control apparatus and bubble traps is necessary to ensure consistency and a high level of particle counter performance. These procedures should be part of the cleaning and maintenance schedules.

Performance verification. Because of their design, particle counters can be sensitive to subtle changes in their optics, electronics, and sampling. It is important to verify the performance of particle counters at a frequency that is stated by the manufacturer. Verification should always be performed at a minimum after any significant maintenance involving cleaning or part replacement. If an instrument fails the manufacturer's recommended verification procedure, the instrument must be recalibrated.

Verification procedures differ depending on the manufacturer of the particle counter. Procedures include using an independent verification standard with a defined number of particles of a known size and matching the instrument response to an independent reference particle counter. Depending on the type of verification, appropriate training should be performed to ensure the techniques and procedures can be correctly executed.

Verification procedures that involve the use of an independent standard material can become expensive, depending on the type of verification and the number of instruments to be verified. However, this is a very important procedure that should be performed to make sure that particle counters are functioning properly. Thus, it is important for the plant to budget adequate resources for operation and maintenance of these instruments.

Calibration. The calibration of particle counters is a complex procedure that will correlate a detector response to a given particle size. The process is typically performed by the manufacturer and can be performed onsite or at the manufacturer's facility. Most manufacturers recommend recalibration on an annual basis. However, if the sensors continue to perform as expected and pass verification, a different schedule can be considered. If the sensor is to be sent to the manufacturer, a replacement strategy should be utilized while the instruments are being serviced.

Practical Applications of Particle Counters for Drinking Water

The size constraints of most light obscuration particle counter flow cells dictate the practical applications of this technology in drinking water. The most common applications in particle counting include the enhanced monitoring of filter effluent, for either conventional or membrane filtration; filter assessment and troubleshooting; and log removal calculations across a filter. Each of these applications will be discussed in more detail.

Data. Online particle counters can produce overwhelming amounts of data in a single filter run. One way to deal with large amounts of data is to determine values for the 10th, 50th, 90th, 95th, and 98th percentiles. Using percentiles to assess data is not difficult if particle count data are entered into a spreadsheet program that can tabulate data in their order of magnitude. Figure 3-7 presents data arranged in this manner. A second approach is to prepare a filter profile using particle counts, showing the cts/mL from the start of the filter run to its conclusion, as is done in Figure 3-8. This approach shows the effect of operating time, while the percentile analysis gives an indication of the degree of variability of particle counts.

More recent approaches in particle counting include data simplification protocols with a backup of more detailed data. Many plants will monitor total counts greater than a given size, which is normally either 2 μm or 3 μm. This protocol typically will be a primary complement to other monitoring trends such as turbidity and changes in operation. Then, if the count trends suddenly increase, additional data can be retrieved if a troubleshooting need exists.

Filter performance. When using particle counters, some general guidance on the types of technology should be acknowledged so that these can be effective monitoring tools. First, particle counters that are best applied are light obscuration and have a size limit of 1 to 2 μm. For membrane filtration, a 1-μm counter will show

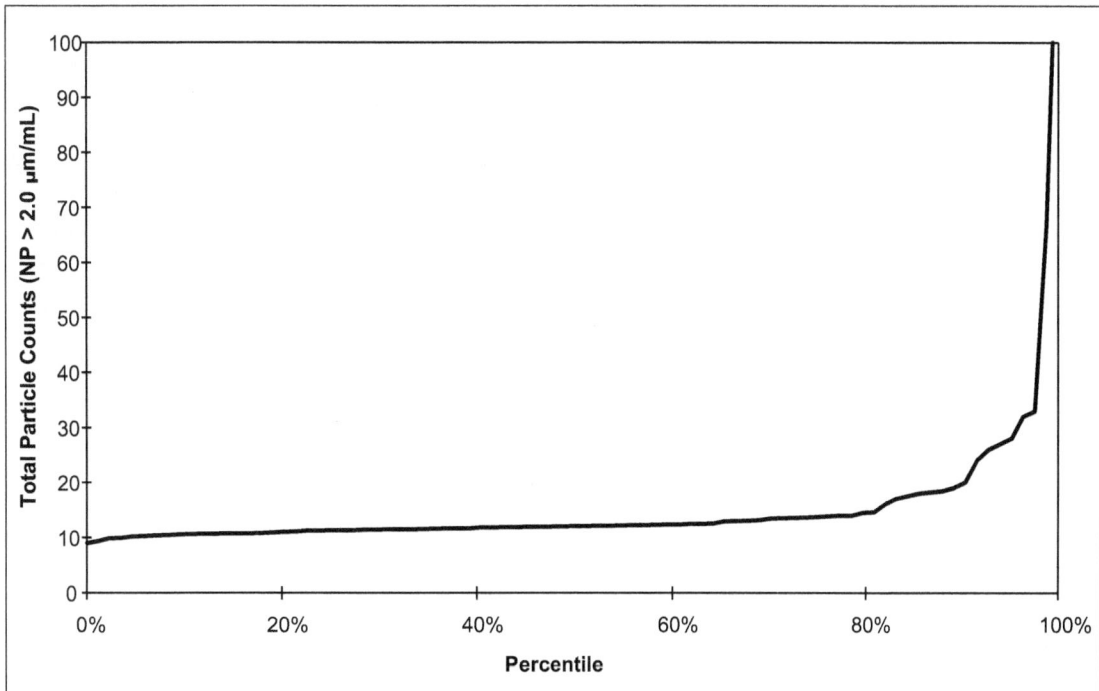

Courtesy of David Pernitsky.

Note: NP = number of particles

Figure 3-7   Filtered water particle count data prepared to provide percentile analysis

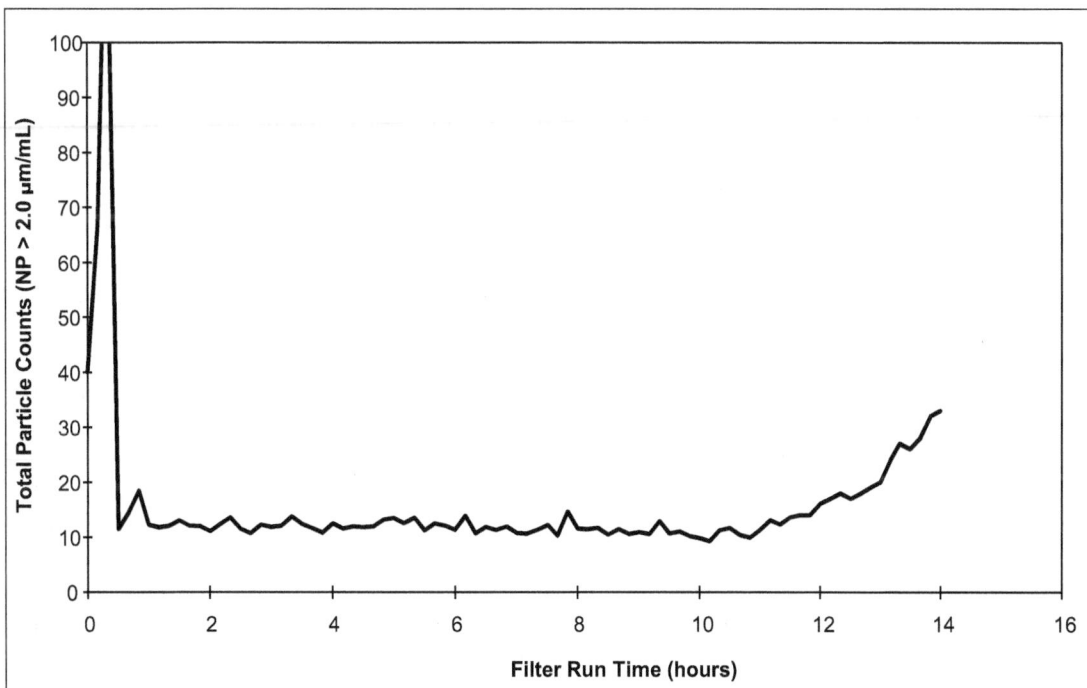

Courtesy of David Pernitsky.

Note: NP = number of particles

Figure 3-8   Particle count data for duration of filter run

distinct advantages with its increased ability to detect fine changes such as single fiber breaches. However, for most conventional filtration applications, the 2-μm particle counters have been shown to be the most successful technologies that can be applied. The 1-μm counters can show advantages in conventional filtration and provide early warning for precursors to filtration breakthrough, but they will warrant more attention and care to sampling and maintenance protocols to ensure that they run effectively and generate quality data.

Particle counters can be very useful at filtration plants where producing very low filtered water turbidity is a goal. Particle counters can prove more effective than turbidimeters for assessing filter performance when filtered water turbidity is 0.10 ntu or lower and small changes in turbidity become more difficult to discern. For example, at a Great Lakes filtration plant particle count data were used to detect a problem that was not easily recognized by analyzing turbidity data. Filters at the plant have two independently operated effluent valves and headers. Flows from each portion of the filter are combined and conveyed to the clearwell. When one filter had slightly elevated particle counts (3 to 5 per mL) in the size range of 2 μm and larger, as compared to other filters (1 to 2 per mL), plant staff investigated and found that one of the two effluent valves had not opened, causing one half of the filter to operate at a rate double that of other filters.

Filter assessment and troubleshooting. Particle counters have been successfully used to assess and/or characterize the performance of filters by indicating the concentrations of various size ranges of particles passing through filters. Counters can also be used gain insight into filtration performance when testing operational changes or when chemical applications such as changing a filter aid chemical are considered. Such studies require a thorough understanding of filtration performance prior to troubleshooting or changing operational protocols. To begin, a particle count and distribution baseline under optimal performance and operation is first performed to generate a particle size and count distribution profile. Then as changes are undertaken in the filtration process, the particle size and count distribution profiles can be generated and compared to the original baseline to determine if the change helps or harms the filtration process.

Log removal. Assessing the removal percentage or "log removal" of particles from raw water to filtered water is problematic in plants that coagulate and filter water. As noted before, the lower threshold for particle counters generally is over 1 μm, but many particles in water are smaller than that size and are not counted. When coagulant is added to raw water, some of the submicron particles not included in the raw water particle count aggregate into larger sizes that can be detected, thus increasing the particle count. However, small countable particles can be coagulated and formed into larger aggregates, thereby decreasing the number of countable particles. With factors working both to increase and to decrease countable particles, assessing the true change in particle numbers through a filtration plant becomes difficult.

The probability of accurate log removal calculations increases if the log removal calculation is performed only across the filter. In this calculation, particle counts at the end of the clarification process can be compared to the filter effluent particle counts. This is because at the end of the clarification process, most of the flocculation process is completed, and if this process is correctly performed, the particle count will not approach the coincidence limits for a given technology. If the calculation is to be performed, it is imperative that the particle count levels be confirmed in the clarifier effluent sample to ensure undercounting is not taking place.

Keep in mind when applying such a technology as particle counting to a flocculation or clarifier sample, a practical limit for particle counters does exist. Excessively high particle counts and large particles can cause operational problems that include coincidence (false negative particle counts) and instrument plugging.

Particle Counting Complements Turbidity

Particle counters are often described as being more sensitive than turbidimeters, which is an easily misinterpreted statement. Turbidimeters are capable of detecting particles as small as 0.01 μm in diameter if sufficient numbers of particles are present to generate a measurable signal on the respective turbidimeter technology (Burlingame et al. 1998). Together, particle counting and turbidity instrumentation can be very complementary. Laser turbidimeters, for example, can detect particle events that are composed of particles that are below the size limitations of a particle counter. Conversely, particle counters can detect low concentrations of large particles that a turbidimeter may not detect. It is important to note that particle counters cannot replace turbidimeters and vice versa. Both have their unique advantages on the ends of the sample size distribution spectrums, and they also provide overlap over other areas of this size distribution spectrum. Thus, they are very complementary technologies.

Limitations in particle counting. Particle counters can size and count particles, but they do have limitations in analytical capability. They are not able to identify the type of particles, such as clay or *Cryptosporidium* oocysts or *Giardia* cysts. Particle counters generally do not detect particles smaller than 1 μm, and often the minimum size detected is 2.5 μm. Particles in water have a variety of shapes, indices of refractions, and optical densities. If the properties of a particle in water are different from the properties of the particles used to calibrate the instrument, the size calculated for the particle may be inaccurate.

Another limitation of particle counters is the difficulty related to comparing data from different brands of these instruments. No standardized design for the sensor exists, whereas for turbidity the nephelometric method has been accepted as the standard. Standard Method 2560 (APHA et al. 2005) has been developed for light blockage particle counters, but as of the date of this publication, no regulatory performance requirement has been developed for analysis of filtered water by particle counting.

A practical consideration for particle counters is related to the size of the instrument's sensor orifice and flow rate. For example, in a counter with a sensor having a particle size range of 2.0 to 400 μm and a concentration limit of 15,000 particles per mL, a typical flow rate is 100 mL/min and a typical orifice size is 1 mm × 1 mm. At this flow rate, the velocity through the orifice is 5.5 ft/sec (1.7 m/sec). *Recommended Standards for Water Works* (Great Lakes–Upper Mississippi River Board of State Public Health and Environmental Managers 2003) suggests 1.5 ft/sec (0.46 m/sec) as the upper limit for the velocity of flocculated water in pipes in a treatment plant. Floc breakage is a concern if particle counters are used to compare clarified water and filtered water at a filtration plant. Floc breakage during counting could also be a problem if particle counters are used to assess clarifier performance by counting flocculated water and clarified water in plants where sweep floc coagulation is practiced, especially if visible floc particles in the size range of 1 mm and larger are seen in the water being sampled for particle counting. Breakage of floc in the counter sensor would indicate presence of a larger number of small particles rather than smaller numbers of large particles and would produce incorrect particle count data.

Summary

Particle counters are a very useful tool that can provide sensitive and descriptive information regarding the particulate content within a sample. In general, particle counters complement turbidity monitoring and provide information in the areas where turbidimeters lack sensitivity, and turbidimeters can provide information where particle counters lack sensitivity. The key to successful use of a particle counter includes understanding the sample composition and potential interferences prior to its application. In applications where particle counting is used, it is important to perform and adhere to scheduled maintenance and performance verification protocols to ensure that quality data are generated that best suit the water treatment plant's needs.

STREAMING CURRENT MONITORS: TECHNICAL DETAILS

Streaming current monitors measure the electrical charge characteristics of particles in water and are used to control coagulation and sludge conditioning processes. These instruments are typically installed online, downstream of coagulant or polymer addition. The postcoagulation particle charge determined by the instrument is used to control chemical feed rate. A change in the streaming current reading away from a predetermined set-point indicates that the coagulant dose is either lower or higher than optimum.

The use of SC technology requires that the operator first identify the optimized condition of coagulation for the plant. Once a coagulant dose is found that results in the production of good quality water in the plant, the streaming current monitor is typically set to a value of zero. Then, small deviations from this baseline can be used as indicators for feedback to dosing. For example, if the instrument has been set to zero and the value deviates in the positive direction, the dosage of coagulant can be reduced to reduce the positive charge in the water. The reverse would occur if the value deviated in the negative direction. Coagulant dose changes are typically done manually by the operator, although in some instances a feedback loop is used to directly control the chemical feed pumps based on streaming current monitor output.

Theory of Operation

A simplified cross section of a streaming current monitor is presented in Figure 3-9. A sample of water flows through the chamber where a small piston moves in a vertically reciprocating motion. A voltage and current are generated as electrically charged particles attached to the surfaces of the piston and the inner-boot move relative to each other. Electrodes in the cylinder sense the voltage/current, and electronic processing generates an output signal.

The piston's motion is sinusoidal, with a typical rate of 3–7 cycles per second. Thus, the resulting current generated is alternating and extremely low, at approximately 10^{-12} amps. To produce a usable and constant signal, the current is amplified, rectified, and filtered prior to being sent to a display or control device. The output is typically displayed nondimensionally, and it is not calibrated to any actual particle charge.

There is no standardized methodology for measuring streaming current, and the magnitude of the current generated in any given instrument depends upon its design, the water chemistry, and the sampling conditions. Therefore, different responses will be seen with different waters and between different instruments. For this reason, most instruments provide either sensitivity or gain adjustment to vary the amplification of the signal to suit the specific application. More information on sensor design and the relationships between streaming current and particle charge can be found elsewhere (Dentel and Kingery 1988).

Factors Affecting Operation

The use of streaming current for coagulation control requires patience. There are interferences that must be considered and mitigated in order to reduce error in its use; this is not an instrument that can simply be plugged in and allowed to run. Rather, the performance of the instrument must be optimized and closely monitored until robust operational and maintenance protocols have been developed for each site. Several of the more common factors affecting operation are discussed in the following sections.

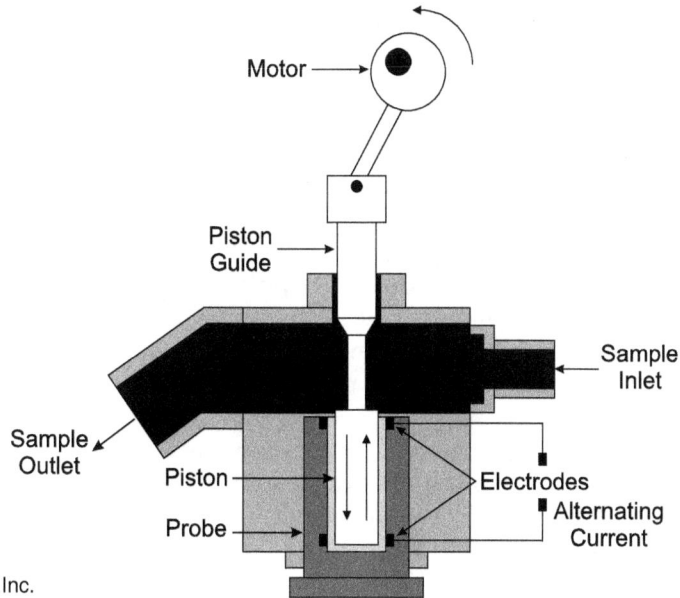

Source: Chemtrac Systems Inc.

Figure 3-9   Schematic diagram of streaming current monitor

Raw water pH.   Any change in pH will affect the streaming current reading, as the surface charge of particles and the charge of the functional groups on NOM molecules are affected by pH. This behavior can be confusing to operators if the coagulant dose has not been changed. High pH tends to accompany higher negative charge for both suspended particles and dissolved NOM molecules.

As well, the positive charge on dissolved aluminum and iron-based coagulant species is less at higher pH (Pernitsky and Edzwald 2003). This means that more coagulant is needed to effect charge neutralization at higher pH compared to low pH, leading to less instrument sensitivity at high pH. Sluggish response and poor sensitivity are often observed when coagulation occurs above pH 7.5 to 8. This lack of responsiveness can be offset somewhat by increasing instrument gain.

Location of sampling point.   The sampling point must be located after thorough mixing to ensure uniform dispersion and reaction of coagulant. Rapid fluctuations in the SC reading (particularly during extremes in process flow) may indicate nonuniform dispersion of coagulants and a need to sample further downstream or a need to move the coagulant addition point further upstream.

The sample point should not be located at a low point in a basin, channel, or pipe. Sand, grit, or other abrasive materials can damage the sensor or clog sample lines. Sampling near the center of a pipe, through a corporation stop diffuser, has been a successful strategy for plants that do not have mechanical rapid mixers or static mixers.

When determining the point at which to sample, consult the manufacturer for guidelines based on experience at similar facilities. It is also advisable to evaluate different sampling points to ensure maximum instrument response with a minimum of signal fluctuation. If possible, perform this evaluation under a variety of operating conditions. Based on field experience, a general guideline is to provide a 1- to 3-min lag time between coagulant addition and when sample reaches the sensor. This period may vary depending on specific plant conditions. The lag time can be intentionally changed in two ways: (1) by selection of the sampling point (where the sampling line draws from the process flow) or (2) by altering the flow time required for the sample to reach the sensor.

Although streaming current monitors can be located upstream of coagulant addition to monitor changes in raw water charge, this approach is not recommended, as it does not provide any information on whether the water has been properly coagulated and conditioned for good clarification and filtration performance.

Coagulant type and strength.   Any change in coagulant type or strength will affect streaming current readings. This should be anticipated, and appropriate modifications to automatic or manual chemical feed rates made to compensate.

Streaming current monitors are often used as indicators of coagulant feed failure, as a lack of coagulant addition or a "bad batch" of coagulant or polymer will result in a rapid change in streaming current.

Determination of proper set-point.   The streaming current set-point is determined by optimizing the plant turbidity/TOC removal and then noting the corresponding streaming current reading. Operational parameters that should be evaluated include filter influent and effluent turbidity/TOC, particle counts, filter run times, head loss buildup, filter rinse volumes, and so on. Most streaming current instruments allow plant personnel to set the streaming current reading to "zero" when optimum conditions are achieved. Overfeed conditions are then indicated as positive readings, while underfeed results in negative readings.

Once the set-point is established, alarm limits can be determined by intentionally causing coagulant underfeed and overfeed conditions. When conditions begin to deteriorate in the plant operations (i.e., filter runs become shorter, filter effluent turbidity levels rise, etc.), the streaming current readings are recorded. These high and low streaming current readings can be used as alarm points to alert plant personnel of potential loss of treatment efficiency.

Calibration and maintenance.   Some treatment chemicals can cause deposition, scaling, and loss of measurement sensitivity. Aluminum and iron coagulants, potassium permanganate, and lime, for example, can cause deposits requiring chemical or mechanical (brush) cleaning. Usually, these foulants cause a slow "drift" over several weeks or months. Between cleanings, if the streaming current instrument has a signal "zero offset," sensitivity can be regained by increasing the electronic gain. A general guideline is to chemically/mechanically clean the sensor components after this gain adjustment is done a couple of times. Eventually, sensors need to be replaced because of physical wear, abrasion, or uncleanable physical deposits on the sensor surfaces. Sensor life can vary between 6 months to 5 years.

No standard calibration procedures have been developed for streaming current monitors. The output of an individual instrument will be dependant on the physical clearances between the piston and cylinder and the electrical properties of the electrodes. Periodic comparison of the operating set point to jar tests or plant performance is adequate for quality control purposes.

Laboratory Streaming Current Analyzers

Recently, bench-top laboratory streaming current units have become available. These units operate on the same principle as online units, but the sensor is immersed in a jar or beaker of coagulated water. Users have found that under extreme and/or rapid turbidity excursions, they are a good supplement to jar tests for defining the best coagulation chemistry.

TOTAL ORGANIC CARBON ANALYZERS: TECHNICAL DETAILS

Online measurements of total organic carbon became popular after the USEPA and other regulators stipulated that water treatment plants achieve specific TOC removals through coagulation. Whereas TOC analysis previously was restricted to laboratory instruments, proven and robust online sensors now can be used to monitor the TOC

Courtesy of David Pernitsky.

Figure 3-10 Typical online TOC instruments

of the raw and treated waters to determine TOC removals quickly and easily. Typical online TOC instruments are shown in Figure 3-10.

Theory of Operation

TOC is an aggregate measurement of the carbon content of dissolved and particulate organic matter present in water and is the most commonly used measure of NOM in drinking water practice. TOC measurements do not provide any information regarding the structure, size, or chemical properties of the NOM being measured.

The sample processing and analysis steps that occur in a typical TOC analyzer are shown in Figure 3-11. First, acid is added to the sample to reduce the pH and convert all of the inorganic carbon (HCO_3^-, CO_3^{2-}, etc.) to CO_2. Next, a nonreactive gas, typically nitrogen, is bubbled through the sample to purge all of the CO_2 from the sample. As an unintended by-product of this step, volatile organics are also purged from the sample matrix. The sample is then buffered, and the remaining organic carbon fraction, the nonpurgeable organic carbon, is oxidized to CO_2. The produced CO_2 is then measured, typically with a nondispersive infrared (NDIR) detector, and expressed as a mg/L concentration of carbon. It should be noted that if the sample is filtered prior to the analyzer (see discussion below), then the online instrument is actually measuring the dissolved organic carbon.

Data handling

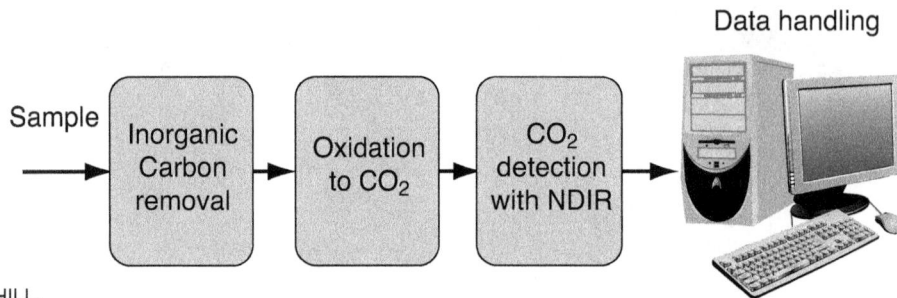

Source: CH2M HILL.

Figure 3-11   Block diagram summary of TOC sample preparation

Two oxidation techniques are used in commercial online TOC analyzers: high-temperature combustion and low-temperature UV-persulfate oxidation. UV-persulfate analyzers are, by far, the more common for drinking water applications.

Low-temperature oxidation involves the addition of sodium persulfate and UV radiation to the sample. Hydroxyl radicals are formed in this process, and these strong oxidizing agents oxidize the organic material to CO_2. Generally, a detection limit of approximately 0.1 mg/L can be achieved with these instruments.

High-temperature analyzers use combustion at temperatures in excess of 680°C, often in the presence of a catalyst, to convert organic compounds to CO_2. This process is very efficient and will allow the detection of some hard-to-oxidize compounds that cannot be detected by the UV-persulfate method. The limit of detection for these instruments is approximately 1 mg/L, however, limiting their usefulness for low-organic waters typical in drinking water treatment applications.

Factors Affecting Operation

Modern online TOC analyzers are designed to operate unattended in remote locations. However, attention should be paid to installation location. These instruments rely on liquid reagents and will require frequent visits by a technician, so the instrument should be located in a clean, dry, air-conditioned environment. A method of disposing of the used reagents, either a sewer connection or a holding tank, will be required.

The location of the sample point for the instrument will depend on the application. Raw water TOC samples should be obtained before the addition of water treatment chemicals. Online TOC analyzers are not designed to handle waters with high turbidity. Depending upon the manufacturer's recommendations, some applications will require the installation of a prefilter upstream of the online analyzer (Figure 3-12). This is especially common on high-turbidity raw waters. Prefiltration may remove a portion of the particulate organic carbon as well as inorganic silts and sediments.

TOC samples intended to quantify TOC removal percentages by coagulation and clarification processes can be taken first in raw water and then downstream of clarification or filtration. Measurements taken downstream of clarification will provide the quickest notification of changes in coagulation performance. However, the presence of floc particles may increase the cleaning and maintenance requirements of the analyzer. As well, postclarification TOC measurements may be higher than measurements taken downstream of filtration, as they will include the TOC that has been coagulated and is contained in the floc particles. Downstream of filtration, the particulate TOC fraction will be very small. A prefilter upstream of the online analyzer can be provided on postclarification installations. Because changes in raw water NOM generally take place over an extended timeframe in natural waters, the additional lag time associated with measuring TOC downstream of filtration is not a concern. For rapidly changing water sources, early warning of increased TOC levels may be best obtained by a raw water sample.

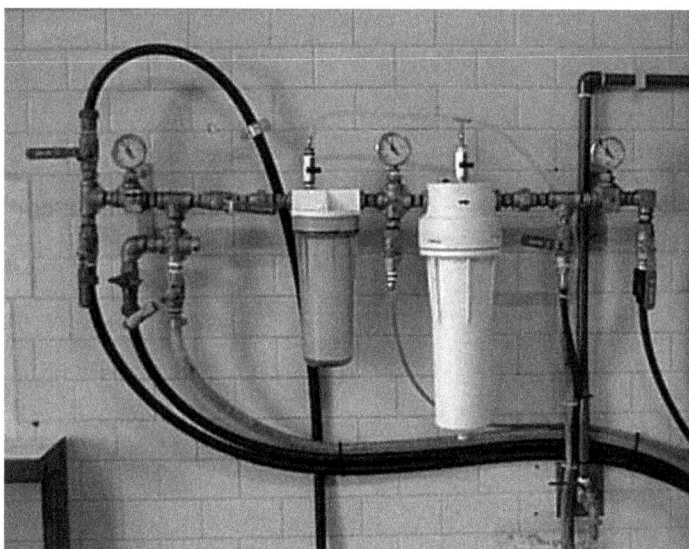

Figure 3-12   Typical TOC prefiltration apparatus

Regardless of the sample location, the instrument should be installed as close as possible to the sample point to minimize travel time. Sample lines should be protected from light to avoid the growth of algae.

Calibration and Maintenance

Online TOC analyzers are relatively complex compared to most instruments in drinking water treatment plants and have above-average calibration and maintenance requirements. Calibration requirements vary by manufacturer, but it is typically recommended to check calibration monthly, after certain maintenance is performed such as UV lamp replacement, or when readings become unstable. Calibration is conducted using standard organic carbon solutions, as described in the instrument manual and *Standard Methods*.

Online TOC analyzer maintenance includes cleaning of the sample handling and reaction chambers; refilling of reagent reservoirs; verification of sample, reagent, and gas flow rates; replacement of peristaltic pump tubing; and replacement of the UV lamp.

ULTRAVIOLET ABSORBANCE/TRANSMITTANCE ANALYZERS: TECHNICAL DETAILS

Measurement of ultraviolet absorbance has been shown to be a useful alternative to TOC measurements for quantifying concentrations of NOM in drinking water treatment (Edzwald et al. 1985). Most organic compounds found in natural waters absorb UV radiation, with carbon–carbon double bonds and ring structures strongly absorbing light at 254 nm. Absorbance and transmittance are therefore useful as measures of the amount of organic matter present in the water for coagulation process control. The exact relationship between UV absorbance and total organic carbon concentration is unique for each raw water source. However, for a given raw water source, increases in UV absorbance indicate increasing NOM concentrations and increasing coagulant demands. Measurements of UV absorbance do not account for all of the organic matter present, as organic matter without rings or double bonds (e.g., simple alcohols) will not be measured, and the analytical procedure requires filtration of the sample, so results are related to the concentration of DOC rather than the concentration of TOC. However, the UV-absorbing fraction represents the fraction of NOM that is most easily removed by coagulation. UV absorbance analyzers are popular with operations staff, as they are typically less expensive and require less maintenance than TOC analyzers.

Because of the importance of carbon–carbon double bonds and ring structures in UV absorbance, persons using UV analyzers need to be aware that oxidants such as chlorine, ozone, and so on can break carbon–carbon double bonds and ring structures in organic compounds. The UV absorbance of oxidized water thus can be lower than the UV absorbance of the water before oxidation, even though no organic carbon was removed by oxidation. This is especially true with the use of ozone. Reduction in UV absorbance with other oxidants such as chlorine, potassium permanganate, and so on will depend on the nature of the NOM and the concentration of the oxidant.

Theory of Operation

The simplest technology for measuring UV absorbance is the single-beam spectrophotometer, four versions of which are shown in Figure 3-13. A light of known wavelength (typically 254 nm) and intensity is passed through a sample cell of known path length. A photodetector on the far side of the sample cell then measures how much light was attenuated by the sample. The difference in the amount of light entering and leaving the sample is reported as the UV absorbance. UV absorbance is an absolute value expressed as absorbance per centimeter. UV absorbance and UV transmittance (UVT) are mathematically related according to Beer's Law.

$$A = -\log_{10}(T) \qquad \text{(Eq 3-1)}$$

Where:
  A is the absorbance (cm^{-1})
  T is the transmittance (expressed as a decimal)
Example: A transmittance of 85 % corresponds to an absorbance of 0.07 cm^{-1}

Frequent recalibration is often required with single-beam instruments, as the output of the source lamp fluctuates with age. To compensate for this, many modern analyzers employ a dual-beam configuration, in which optics are used to split the source light into two beams. One beam is directed through the sample and the other beam is used as a reference. The amount of light absorbed by a sample depends on the path length through the sample cell; larger cells will have a longer path length, and therefore more light will be absorbed. Larger path lengths are required for low-organic

Trojan Optiview	**HF Scientific AccUView**	**ChemScan UV-0254**	**Wedeco TMO-III**
Source: Trojan UV.	*Source:* HF Scientific.	*Source:* Applied Spectrometry Associates Inc.	*Source:* ITT Water & Wastewater USA.

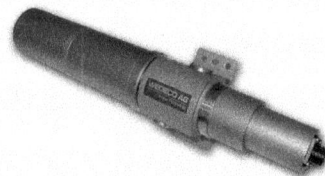

Figure 3-13   Examples of single-beam UV absorbance analyzers

waters. Wastewater applications, conversely, will require small path lengths, because of the high concentration of organics present. For most online instruments, sample path lengths vary between 1 and 10 cm. In North America, by convention, results are displayed as absorbance per cm (cm^{-1}).

Advanced UV absorbance analyzers that use diode array technology to scan the multiple wavelengths in the ultraviolet-visible light spectrum (up to 256 wavelengths) are available to enable measurement of other parameters to provide more detailed characterization of NOM, as well as to measure nitrate/nitrite, phosphates, free and total chlorine, some metals, turbidity, and disinfection by-products.

Factors Affecting Operation

UV analyzers are inexpensive, robust, and simple to operate. Unlike TOC analyzers, UV analyzers do not rely on chemical reagents. Because these instruments use optical methods, the results are instantaneous, and the flow from the instrument can be reintroduced into the main flow or disposed of in a sewer.

Suspended particles can scatter light, and therefore interfere with absorbance measurements, so prefiltration is necessary upstream of UV absorbance analyzers for all but filtered water samples. A typical filter apparatus is shown in Figure 3-12. Alternatively, some analyzers measure absorbance at a second wavelength for turbidity compensation purposes.

Fouling of the optical sample windows can also result in errors. Inorganic precipitation and biofilms can obscure the sample cell, leading to unexpectedly high absorbance measurements. Some instruments come with automated wiper mechanisms to remove foulants. Similarly, the presence of other UV-absorbing substances such as dissolved iron, nitrate, and nitrite can interfere with measurements of NOM.

Calibration and Maintenance

Absorbance instruments are typically calibrated at the factory, and primary calibration is rarely done in the field. Some manufacturers include an optical filter that can be used to verify the calibration of the instrument. If this calibration verification deviates too far from the manufacturer's specifications, it is usually recommended that the instrument be returned to the factory for service. However, if proper quality assurance procedures are followed, skilled laboratory technicians can calibrate these instruments against known standard organic carbon solutions, as described in *Standard Methods*.

Typical quality assurance checks done in the plant include checking the baseline absorbance of an organic-free blank solution and comparing absorbance readings against a well-maintained laboratory spectrophotometer. Most calibration errors are the result of dirty sample cells or optics.

Maintenance requirements include replacing the lamp according to the manufacturer's recommendations, cleaning the optical surfaces of the sample cell, and replacing any upstream filters as necessary.

ONLINE MONITORING OF pH: TECHNICAL DETAILS

Measuring pH is fundamental to water treatment operations, so some utilities have elected to use online monitoring to ease the operator's burden related to measuring pH. In the *Filter Maintenance and Operations Guidance Manual*, Logsdon et al. (2002) summarized information provided by 37 plants that participated in the project and reported on their procedures for management of coagulation and filtration. Of the 37, 10 used online pH monitoring for settled water, and 17 monitored pH of combined filter effluent continuously. Five plants treating surface water by lime softening reported on monitoring practices, and of the five, three measured settled water pH online, while two measured combined filtered water pH online.

Theory of Operation

In simple terms, the measurement of pH is the measurement of the concentration of hydrogen ions (H^+) in a water sample. Pure water consists of molecules of H_2O. A very small fraction of these molecules in pure water dissociate, forming H^+ and OH^- (hydroxyl) ions. In acidic solutions (pH less than 7), the concentration of hydrogen ions in solution is greater than the concentration of hydroxyl ions. In basic solutions (pH greater than 7), the opposite is true. At neutral pH (pH = 7), the concentrations of hydrogen and hydroxyl ions are equal.

For any given temperature and ionic strength, the product of the hydrogen and hydroxyl ions is a constant. At 25°C in pure water, the equilibrium for hydrogen and hydroxyl ions is expressed as:

$$[H^+][OH^-] = K_w = 1.01 \times 10^{-14} \text{ where } K_w \text{ is the ion product of water}$$

At pH 7, therefore, $[H^+] = [OH^-] = 1.005 \times 10^{-7}$, and the concentrations of $[H^+]$ and $[OH^-]$ are each approximately 0.0000001 mol/L or 10^{-7} mol/L.

Technically, the terms $[H^+]$ and $[OH^-]$ represent the activity of hydrogen ions and hydroxyl ions in mol/L. When the ionic strength is less than 0.1 mol/L, the molar activity is approximately the same as the molar concentration. For brackish or seawater applications, activity corrections are necessary.

In water treatment applications, hydrogen ion concentration is expressed in terms of pH, which is defined as the negative log of the hydrogen ion activity (or concentration, when dealing with drinking water, as discussed above).

$$pH = -\log[H^+]$$

For example, when the concentrations of hydrogen and hydroxide ions are equal (neutral pH) the hydrogen ion concentration (activity) is 1.005×10^{-7} at 25°C. The log to base 10 for the activity of 1.005×10^{-7} is -7, so the pH is 7.

According to Method 4500-H^+ in *Standard Methods for the Examination of Water and Wastewater* (APHA et al. 2005), electrometric pH measurement involves determination of hydrogen ion activity by potentiometric measurement using a reference electrode and a glass electrode, in which the electromotive force produced in the glass electrode system varies in a linear fashion with pH. The pH electrode consists of two types of glass, a stem of nonconducting glass and a tip, often bubble shaped, that is made of a pH-sensitive lithium ion conductive glass. Lithium ion electrons can be exchanged across the conductive glass by hydrogen ions, and this creates the millivolt potential. Temperature influences the instrument reading, so temperature compensation is often provided to account for this.

At high pH values, sodium ions can interfere with pH measurement. *Standard Methods* recommends use of a "low sodium error" electrode for measuring pH over 10, because of the possibility that standard glass electrodes can produce pH readings that are too low. At water treatment plants practicing coagulation (not lime softening), the pH during treatment should not reach 10, so sodium ion interference would not be expected to occur.

Factors Affecting Operation

Online measurement of pH at coagulation plants is especially helpful if the source water quality can change rapidly. This can be encountered in flashy streams and rivers that are subject to rapid increases in flow caused by runoff from heavy precipitation. Another situation that can result in frequent changes of pH relates to water bodies that have diurnal patterns of algae growth. As algae grow during daytime when photosynthesis is occurring, they take up carbon dioxide, and decreasing the concentration of carbon dioxide raises the source water pH. Algae growth and uptake of carbon dioxide cease at night. When daily pH changes are expected to happen, using an online measurement takes an important analytical burden off the operating staff. Online monitoring is especially valuable if the treatment goals included optimizing particle removal or TOC removal by coagulation and filtration.

Budd et al. (2004) discussed the importance of pH in coagulation process control, and Edzwald and Kaminski (2007) discussed the use of metal coagulants at the pH of minimum solubility. For water utilities that must comply with regulatory requirements for reduction of the concentration of TOC in water, pH again is an important water quality measurement.

The pH data obtained by online monitoring are often used for multiple purposes. If online pH data for coagulated water are also used in assessing CT values for chlorine disinfection, for which pH must be known, then calibration of the pH monitoring instrument may need to be done in accordance with regulatory agency requirements, and it may be necessary to keep records of instrument calibration for future regulatory reference.

Calibration and Maintenance

Control of pH is necessary for optimal treatment plant performance. At plants where pH is especially critical, such as for lime softening, for optimum removal of particles or TOC, or for frequent calculation of CT values related to chlorination, online monitoring is very helpful to plant operations. If, however, the online pH data are inaccurate, the result could be failure to properly stabilize lime-softened water resulting in precipitation of calcium carbonate in filter beds or water mains, inadequate removal of TOC or particles, or collection of invalid data related to chlorination efficacy.

The manual provided by the instrument manufacturer should be consulted for instructions on calibration procedures and the recommended frequency for calibration. To attain valid pH data, plant staff must perform calibration and other maintenance procedures at least as frequently as specified in the instrument manual. Eventually, all pH probes must be replaced. Typical lifespan of a pH probe is 6 months to 2 years.

Records of calibration procedures should be kept as a means of assessing the continued usefulness of the instrument and the data it provides. Furthermore, if online pH measurement is used to provide data related to evaluation of CT values, it is highly advisable to keep accurate records of instrument calibration even if keeping such records is not a regulatory requirement, and it is essential to ensure that the instrument uses a method approved by the USEPA.

Online pH measuring technology was reviewed by Haught et al. (2002), who described the performance and features of several different types of online pH probes. Information provided by the authors included range of pH measured, accuracy, calibration requirements, and general comments about the probes, such as automatic temperature compensation. These aspects of pH probes should be considered when purchasing an online instrument.

Electrometric pH measurement should be less influenced by instrument design than measurement of turbidity or counting of particles in water, as online and bench-scale instruments share a fundamentally common design. Therefore, the use of bench-scale pH meters to periodically check the validity of online pH data is encouraged as a means of having confidence in the online pH data.

However, it should be noted that an online pH electrode is in continuous contact with the sample stream and is therefore conditioned to the temperature and ionic strength of the process stream. The pH probe of a laboratory instrument is typically stored at room temperature in a buffer solution. These differences can result in sluggish performance and/or measurement discrepancies, especially in cold waters.

One precaution with comparing pH data from laboratory instruments and online instruments relates to waters for which pH may be influenced by exposure to the atmosphere. An in-line pH electrode does not measure samples in an open container, so atmospheric exposure would not be a problem. Exposure to the atmosphere during grab sampling for laboratory measurements can introduce measurement errors. If pH is changed in a closed system such as a pipe in a treatment plant and if, as a result of the pH change, the water in the pipe is no longer in an equilibrium state with the atmosphere, gain or loss of carbon dioxide during a pH measurement made in an open system could result in a different pH reading from a laboratory instrument versus one made online in the closed system (pipe). Evidence of this problem may be exhibited as "drifting," in which the laboratory pH meter readout rather quickly seems to reach a value but then it gradually changes with time. This "drifting" has been observed in performance of jar tests with low-alkalinity water in which pH changes have been made to evaluate the appropriate pH for coagulation. Schock et al. (1980) published a procedure for measuring pH in a closed system for samples that may exhibit a pH change due to exposure to atmospheric carbon dioxide.

Another potential problem for online pH electrodes is the possibility of fouling of the glass electrode by precipitates. Metal oxides, such as oxides of manganese or iron, can coat the electrode and interfere with ion transport. If a pH electrode is continuously exposed to water treated with potassium permanganate, this may be a condition to watch for. Another common cause of fouling is precipitation of calcium carbonate on an electrode at a lime softening plant. These precipitates can result in sluggish response and measurement errors.

The bottom line for online pH measurement is that the data provided by remote monitoring are good only if the instrument is functioning properly and is properly calibrated. Making the assumption that instrument readings are good without verifying this periodically by checking with a laboratory instrument is risky. The importance of pH is such that risks associated with operating a plant at the wrong pH can be high, and failure to ensure the quality of pH readings can be costly in terms of precipitation of calcium carbonate on filter media or pipe walls or in terms of poor coagulation and filtration or inadequate disinfection. Online pH measurement can be a great time-saver and aid to attaining excellent water treatment when the data obtained in this manner are valid, so the payoff for maintaining a good program of quality control can be quite substantial.

REFERENCES

American Public Health Association (APHA), American Water Works Association (AWWA), Water Environment Federation (WEF). 2005. *Standard Methods for the Examination of Water and Wastewater*. 21st ed. Washington, D.C.: APHA.

ASTM International. 2007. *Standard Test Method for the Online Measurement of Turbidity Below 5 NTU in Water*. West Conshohocken, Pa.: ASTM International.

Bellamy, W.D., J.L. Cleasby, G.S. Logsdon, and M.J. Allen. 1993. Assessing Water Treatment Plant Performance. *Jour. AWWA*, 85(12):34–38.

Budd, G.C., A.F. Hess, H. Shorney-Darby, J.J. Neemann, C.M. Spencer, J.D. Bellamy, and P.H. Hargette. 2004. Coagulation Applications for New Treatment Goals. *Jour. AWWA*, 96(2):102–113.

Burlingame, G., M. Pickel, and J. Roman. 1998. Practical Applications of Turbidity Monitoring. *Jour. AWWA*, 90(8):57–69.

Dentel, S., and K. Kingery. 1988. *An Evaluation of Streaming Current Detectors*. Denver, Colo.: AWWA and Awwa Research Foundation.

Edzwald, J.K. 1993. Coagulation in Drinking Water Treatment: Particles, Organics, and Coagulants. *Wat. Sci. Technol.*, 27(11):21–35.

Edzwald, J.K., W.C. Becker, and K.L. Wattier. 1985. Surrogate Parameters for Monitoring Organic Matter and THM Precursors. *Jour. AWWA*, 77(4):122–132.

Edzwald, J.K., and G.S. Kaminski. 2007. A Simple Method for Water Plant Optimization and Operation of Coagulation. *Proc. 2007 AWWA Water Quality Technology Conference*, Session W3. Denver, Colo.: AWWA.

Emelko, M., P. Huck, I. Douglas, and J. van den Oever. 2000. *Cryptosporidium* and Microsphere Removal During Low Turbidity End of Run and Early Breakthrough Filtration. *Proceedings of the 2000 WQTC*. Denver, Colo.: AWWA.

Great Lakes–Upper Mississippi River Board of State Public Health and Environmental Managers. 2003. *Recommended Standards for Water Works*. Albany, N.Y.: Health Education Services.

Hach Company. 1996. *Method 8195, Determination of Turbidity by Nephelometry—USEPA Accepted Method*. Loveland, Colo.: Hach.

Hargesheimer, E.E., and C.M. Lewis. 1995. *A Practical Guide to Online Particle Counting*. Denver, Colo.: Awwa Research Founation and AWWA.

Hargesheimer, E., O. Conio, and J. Popovicova, eds. 2002. *Online Monitoring for Drinking Water Utilities*. Denver Colo.: Awwa Research Foundation and CRS PROAQUA.

Haught, R., and M. Fabris (authors), with E. Hargesheimer (contributor). 2002. Inorganic Monitors. In *Online Monitoring for Drinking Water Utilities*. E. Hargesheimer, O. Conio, and J. Popovicova, eds. Denver, Colo.: Awwa Research Foundation and CRS PROAQUA.

Huck, P., B. Coffey, C. O'Melia, and M. Emelko. 2000. Removal of *Cryptosporidium* by Filtration Under Conditions of Process Challenge. *Proceedings of the 2000 WQTC*. Denver, Colo.: AWWA.

International Organization for Standardization (ISO). 1999. *Water Quality—Determination of Turbidity*. ISO 7027. Geneva, Switzerland: ISO.

Kingsbury, F.B., C.P. Clark, G. Williams, and A.L. Post. 1926. The Rapid Determination of Albumin in Urine. *J. Lab. Clin. Med.,* 11:981.

Logsdon, G.S, A.F. Hess, M.J. Chipps, and A.J. Rachwal. 2002. *Filter Maintenance and Operations Guidance Manual.* Denver, Colo.: Awwa Research Foundation and AWWA.

Pernitsky, D.J., and J.K. Edzwald. 2003. Solubility of Polyaluminum Coagulants. *Jour. Water Supply: Research and Technology–AQUA,* 52(6):395–406.

Pernitsky, D.J., and J.K. Edzwald. 2006. Selection of Alum and Polyaluminum Coagulants: Principles and Applications. *Jour. Water Supply: Research and Technology–AQUA,* 55(2):121–141.

Pernitsky, D.J, and L. Meucci. 2002. Physical Monitors. In *Online Monitoring for Drinking Water Utilities,* Cooperative Research Report, E. Hargesheimer, O. Conio, and J. Popovicova, eds., 63–132. Denver, Colo.: Awwa Research Foundation and CRS PROAQUA.

Sadar, M. 1998. *Turbidity Science Technical Information Series.* Loveland, Colo.: Hach.

Sadar, M. 1999. *Turbidity Standards—Technical Information Series.* Loveland, Colo.: Hach.

Sadar, M., L. Oxenford., M. Lictwardt, and F. Watt. 2003. Monitoring Membrane Integrity Across Four Micro Filtration Pilot Plants Using Multiplexed Light Scattering Technologies. *Proceedings of the 2003 WQTC.* Denver, Colo.: AWWA.

Schock, M. R., W. Mueller, and R. W. Buelow. 1980. Laboratory Technique for Measurement of pH for Corrosion Control Studies and Water Not in Equilibrium With the Atmosphere. *Jour. AWWA,* 72(5):304–306.

US Environmental Protection Agency (USEPA). 1993. Determination of Turbidity by Nephelometry—Method 180.1. In *Methods for the Determination of Inorganic Substances in Environmental Samples.* EPA/600/R-93/100. Washington, D.C.: USEPA.

USEPA. 1997. 40 CFR Parts 122, 136, et al. *Guidelines Establishing Test Procedures for the Analysis of Pollutants Under the Clean Water Act; National Primary Drinking Water Regulations; and National Secondary Drinking Water Regulations; Analysis and Sampling Procedures; Final Rule. Fed. Reg.* 72:47. Washington, D.C.: USEPA.

USEPA. 1998a. 40 CFR Parts 9, 141, and 142. *National Primary Drinking Water Regulations: Interim Enhanced Surface Water Treatment; Final Rule. Fed. Reg.* 63:241:69478–69521. Washington, D.C.: USEPA.

USEPA, 1998b. 40 CFR Parts 9, 141, and 142. *National Primary Drinking Water Regulations: Disinfectants and Disinfection By-products; Final Rule. Fed. Reg.* 63:241:69390–69476. Washington, D.C.: USEPA.

USEPA. 1999a. *Enhanced Coagulation and Enhanced Precipitative Softening Guidance Manual.* EPA 815-R-99-012. Washington, D.C.: USEPA.

USEPA. 1999b. *Guidance Manual for Compliance With the Interim Enhanced Surface Water Treatment Rule: Turbidity Provisions.* EPA 815-R-99-010. Washington, D.C.: USEPA.

USEPA. 2002. 40 CFR Part 141. *Unregulated Contaminant Monitoring Regulation: Approval of Analytical Method for Aeromonas; National Primary and Secondary Drinking Water Regulations: Approval of Analytical Methods for Chemical and Microbiological Contaminants. Fed Reg.* 67:209:65888–65902. Washington, D.C.: USEPA.

Whipple, G.C., and D.D. Jackson. 1900. A Comparative Study of the Methods Used for the Measurement of the Turbidity of Water. *Massachusetts Institute of Technology Quarterly,* 13:274.

Chapter **4**

Flocculation and Clarification Processes

George Budd, James Farmerie, and Paul Hargette

INTRODUCTION

The operation of a coagulant-based water treatment sequence relies on two fundamental sets of processes irrespective of whether the treatment sequence applies conventional gravity-based settling, newer high rate clarification concepts, or a direct filtration sequence in which no clarification step is applied. The first set of processes consists of the mixing and flocculation steps, which include the addition of a coagulant chemical and application of mixing to disperse the chemical, followed by conditions under which the chemically conditioned particles will come into contact to build floc of appropriate size for removal in downstream particle removal processes. The second set of processes consists of the downstream clarification and filtration steps in which the flocculated particles are separated from the water.

The first section of this chapter focuses on the mixing and flocculation steps, which are the first set of processes in the overall sequence. Proper performance at this stage is critical to overall plant performance. In fact, a failure to achieve desired performance in the subsequent clarification and filtration steps can often be traced to poor performance of these processes. The second section of this chapter focuses on the clarification step. The filtration step is covered in chapter 5.

Specific Mixing and Flocculation Goals

Specific goals for mixing and flocculation processes are characterized by different objectives for each process. In the case of mixing, the objective is a rapid dispersion of the coagulant chemical that provides relatively uniform distribution to the particles.

This ensures the most effective use of the coagulant for chemically conditioning all of the particles.

Once the coagulant chemical has been dispersed, flocculation is initiated to provide mixing energy that promotes contact and aggregation of the chemically conditioned particles into floc that can be removed by clarification and filtration. The specific goal for this process step is a controlled introduction of hydraulic shear that causes contact between particles, creating a progressive increase in floc size. Ideally, this process will result in a uniform floc size with particles in the best size range for ready separation.

The attributes of the flocculation step are dictated by the desired floc characteristics for the downstream clarification/filtration sequence. As an example, dissolved air flotation (DAF) targets smaller floc sizes than conventional sedimentation because of the requirement that the floc attach to air bubbles and float. While this comparison represents an extreme, other high rate processes are optimized under different floc size distributions as well. Direct filtration is a special case in which a filterable floc size is the target. In this case, large flocs may cause filter blinding. Even in the case of conventional settling, there is usually a maximum desirable floc size because of the less dense nature and susceptibility to shear that is often encountered with large floc. Therefore, it is important that the particle size objectives be understood in configuring and operating the flocculation step within a plant.

Velocity Gradient in Mixing and Flocculation

The concept of the velocity gradient is important for both rapid mixing and flocculation. The intensity of mixing is generally quantified by the velocity gradient, which is often expressed as G with units of s^{-1} (seconds to the minus 1 power). The velocity gradient is calculated using the energy dissipation rate in the fluid. A higher number indicates more intense mixing.

Velocity gradient varies significantly with water temperature (viscosity); therefore, flocculator speed may need to be changed seasonally where significant changes in raw water temperatures occur throughout the year. When the optimum velocity gradients have been identified by jar testing as described in chapter 2, the plant mixer(s) and flocculator speed/power can be set accordingly by adjusting the unit's variable frequency drive (VFD) speed up or down to match the power indicated by the following formula:

$$P \text{ (kW)} = (G^2 \mu V)/0.7/1{,}000$$
or
$$P \text{ (hp)} = 1.341(G^2 \mu V)/0.7/1{,}000$$

Where:

   0.7 = nominal overall loss efficiency with the mixer VFD, motor, and power train

   μ =   dynamic viscosity in Ns/m^2 (e.g., when water temperature = 40°F (4°C), μ = 0.001569; when water temperature = 50°F (10°C), μ = 0.001308 and when water temperature = 60°F (16°C), μ = 0.001109)

   V =   specific mixing chamber volume in m^3

   P =   power (kW or hp) applied to the mixer motor—usually read directly at the motor VFD display or calculated from amperes displayed at the drive ammeter where speed adjustment is by mechanical speed variators rather than VFD speed.

A procedure recommended by D.K. Nix (2009, pers. comm.) for quick calculation of G in full-scale basins or mixing chambers is:

   G = $388\,P^{0.5}$

Where P is the horsepower per 1,000 gal (3,785 L) of volume in the basin or chamber being mixed. This formula applies when the water temperature is 50°F (10°C).

To calculate horsepower per 1,000 gal (3,785 L) for a basin with one mixer, divide horsepower of the mixer or flocculator motor by the volume of the basin in gallons and multiply that result by 1,000. For a baffled flocculation basin with three separate chambers, each with a motorized flocculator, the G value of each chamber must be calculated separately using the horsepower of the motor driving the flocculator in the chamber and the volume of that chamber. This formula is an approximation, as it does not account for energy (power) losses in the motor or in the drive mechanism.

Viscosity of water varies with temperature, so the multiplier factor must be changed for other water temperatures. Table 4-1 presents factors for water temperatures ranging from 39.2°F to 70°F (4°C to 21.1°C).

It should be noted that, while the optimum G value developed in a jar test provides valuable information and a general indication of proper mixing, the optimum full-scale G values may be somewhat different because of scale-up issues (Amirtharajah et al. 1991). The relationship between optimum jar test G values and full-scale G values varies based on a number of factors and is often best understood based on site-specific experience. Flocculation intensity is discussed further later in this chapter.

RAPID MIX

Rapid-mix concepts have evolved substantially in recent years in recognition of the need for efficient dispersion of chemical as an initial step of coagulant application. This is based on both field observation and research into this aspect of treatment, which indicate a need for distribution of the coagulant within the shortest possible time (Amirtharajah et al. 1991, Clark et al. 1994). Older rapid-mix basins that were designed for detention times of 1 min or more as required under some criteria are not as efficient in accomplishing this goal as compared with newer concepts that tend to be based on short detention time and rapid dispersion methods. These dispersion-oriented configurations can include basins in which mixing is confined to the smallest possible volume as flow passes the mixer blade, and also a variety of in-pipe configurations that include mechanical mixers, static mixers, and hydraulic jet dispersion concepts. Even simple in-pipe addition can be more effective for rapid dispersion than older rapid mix basins that are designed based on detention time, because the confined cross section at the point of application is more amenable to dispersion (Clark et al. 1994).

Operational Considerations

The main operational consideration for the rapid-mix portion of a water treatment facility is ensuring that the correct amount of chemicals is being added based on raw water quality to achieve the proper coagulant dosage and coagulation pH, the two fundamental

Table 4-1   Multiplier factors to convert horsepower/1,000 gal to G at various temperatures

Temperature, °F (°C)	Factor
39.2 (4)	354
50 (10)	388
60 (15.6)	418
70 (21.1)	448

parameters for coagulation. As indicated above, true rapid mixing occurs over a very short time period. Addition of a coagulant chemical (alum, ferric chloride or sulfate, poly-aluminum chloride, etc.) is essential and is sometimes accompanied by the addition of a pH-adjustment chemical (sodium hydroxide, lime, sodium bicarbonate, etc.), depending on the raw water characteristics. Traditionally, when an alkaline pH adjustment chemical is needed to provide alkalinity, practice has been to add the pH-adjustment chemical prior to the coagulant, so alkalinity is available to react with the coagulant. In some instances, reversing this order of chemical addition may provide some treatment benefit. Figure 4-1 provides an example of jar testing that evaluated the effect of the sequence of chemical addition on settled water turbidity. These results indicated an improvement in settled water turbidity with the addition of alum prior to caustic. This improvement was seen in cold water testing but was not evident in subsequent warm water testing.

While not as common, the addition of an acid to reduce the pH of coagulation can also be a consideration. This practice has increased in recent years as reduced pH can lead to increased removal of total organic carbon (TOC) to meet goals for enhanced coagulation and also can help to reduce aluminum residuals when the coagulant is aluminum-based.

While the older, detention-based rapid-mix basin is not an effective concept for coagulant dispersion, it may serve as a high-intensity, short-detention flocculation step that plays a role in the formation of seed floc. This is thought to provide a benefit in some situations (Janssens 1992). Figure 4-2 shows the effect of varying this aspect of treatment in a bench-scale test that was calibrated to match the performance of a plant that was experiencing difficulties with cold-water coagulation. In this case, provision of an appropriate time of rapid mix was an important consideration for calibrating the bench-scale results as shown here. Observations at a number of locations tend to indicate varying levels of effects from this stage of the process. Where such rapid-mix basins exist, it is sometimes beneficial to retain them. In some cases, relocating the point of coagulant addition to an upstream in-pipe location can provide improved dispersion while preserving the rapid-mix basin as an initial flocculation step.

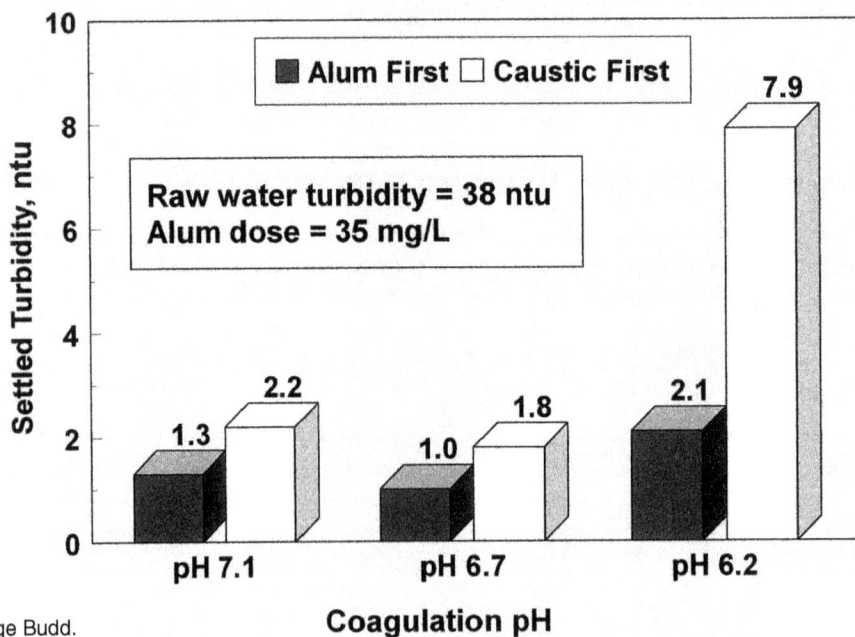

Courtesy of George Budd.

Figure 4-1   Effect of sequence of chemical addition on alum coagulation

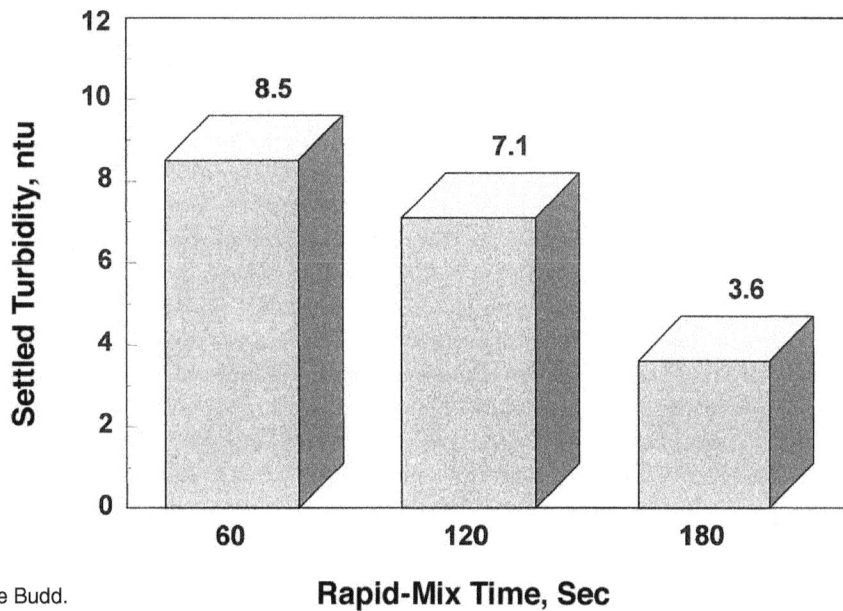

Courtesy of George Budd.

Figure 4-2   Effect of rapid-mixing time on settled turbidity (same flocculation and sedimentation conditions for each mixing condition)

FLOCCULATION

Flocculation configurations generally tend to fall into two general categories: (1) mechanical mixing methods and (2) hydraulic mixing methods, which utilize baffling or other methods for inducing turbulence in a flowing stream. The main advantage of mechanical mixing is that energy input into the system is independent of the rate of flow. The tendency in recent years has been to apply mechanical mixing methods to provide greater flexibility for adjusting flocculation speed to match variations in both water quality and flow rate, but some hydraulic-baffle-type flocculators provide good service and have the potential to provide both simplicity and economy of operation, because energy input is attained through head loss during flow rather than by motor-driven mixing devices.

Mechanical Mixing

The most common types of mechanical mixing equipment are vertical turbine and paddle-wheel flocculators. Walking-beam flocculators are also present at a number of older facilities but are not often installed in new facilities.

Vertical turbine flocculators.   Many of the recent innovations in flocculation have occurred with newer vertical turbine equipment through advances in the application of axial-flow, foil-type impellers that reduce the amount of torque required for a given level of flocculation intensity (Clark et al. 1994). This can simplify the overall mechanics of equipment and reduce its size. The treatment result with this equipment is also good when properly configured with multiple cells that are operated in series, appropriate impeller diameter, impeller placement at a depth that yields the most effective cell-wide agitation, and appropriate cell geometry.

Vertical turbine flocculators have a significant operation and maintenance benefit when all of the serviceable moving parts are placed above the water surface, as is typical of designs in recent years. This advantage, however, may be offset to an extent in larger plants because a large number of flocculators can be required. This derives from the fact that the size of individual flocculation cells can be limited by the need to maintain the surface dimensions in relative proportion to a reasonable overall basin depth, thereby increasing the number of cells required as plant size increases.

Paddle-wheel flocculators. Paddle-wheel flocculators can be provided in both horizontally and vertically mounted configurations. The process performance of this equipment varies widely and tends to be sensitive to flow pattern through the basin, but a high level of performance can be achieved when the flow pattern approximates a plug flow condition. This can be achieved when horizontally mounted paddle flocculators are installed in a basin with a serpentine baffling pattern that directs flow along the shaft of the paddle. Lower process effectiveness is observed in other configurations such as a horizontally mounted flocculator in which flow is perpendicular to the shaft and in vertically mounted paddle flocculators. This reduction in performance is to the result of a greater tendency toward short-circuiting in these basins; the most significant deterioration occurs with flow across a horizontal flocculator.

It should be noted that while the operational, mechanical, and potential process advantages of vertical turbine flocculators have increased the popularity of these devices in recent years, a properly configured horizontal flocculator with a plug flow character may provide slightly higher flocculation efficiency (Clark et al. 1994). This configuration can be easily arranged through the use of serpentine baffling to provide a plug flow characteristic that yields the greatest control over the floc size distribution. Floc size distribution, in particular maintaining a uniform floc size, can be important for optimum operation of downstream clarification processes, particularly for high-rate plate settlers. While the submerged bearings associated with paddle flocculators are not as easily maintained as the above-water assemblies of vertical turbine flocculators, newer bearing assemblies have become available to replace the older metallic roller bearings. In the case of the drive units, direct drive systems have become more widely available as a replacement for the older chain and sprocket type.

Hydraulic Mixing

Hydraulic mixing is typically accomplished utilizing either horizontally baffled (around-the-end) channels or vertically baffled (over- and underflow) channels. The main advantages of these types of systems are that there are no moving parts and no direct power usage. The main disadvantages are that there can be significant head loss in the baffled basins (as much as 2 to 3 ft or 0.6 to 0.9 m) and mixing intensity is a function of plant flow rate, meaning that low mixing intensity is provided at low flow rates, and floc shear may be a problem at high flow rates. However, experience has shown that good flocculation often occurs when Gt (velocity gradient multiplied by time of mixing of flocculation) values range from 3×10^4 to 2×10^5, meaning that the low intensities at low flow rates can work acceptably if available detention times provide Gt values within the indicated range (Kawamura 2000).

Operational Considerations

There are a number of operational considerations for full-scale flocculation basins:

Detention time attributes. Detention time attributes are affected by basin volume, baffling, and staging of flocculation cells. Baffling and staging provide control of short-circuiting to achieve the most effective use of basin volume. The significance of these effects is illustrated in a baffling modification that was made to one of the water treatment plants for the city of Fort Collins, Colo. (Foellmi and Bryant 1991). The original configuration, as shown in Figure 4-3, yielded a short flow path across each flocculation stage that resulted in significant short-circuiting. This configuration was modified by altering the baffling configuration to provide flow along the long axis of each flocculator, as shown in Figure 4-4. This change provided a much more uniform floc size and substantially altered the flocculation efficiency such that the average settled water turbidity was lowered from 1.5 ntu during testing of the initial configuration to 0.7 ntu following the baffling modification. This plant treats raw water with a relatively low turbidity that is less than 10 ntu. A settled water turbidity of 1.0 ntu or less was targeted for meeting overall treatment goals that include filtered turbidities that are consistently less than 0.1 ntu. This modification was critical for meeting overall goals at the facility. Ways to assess short-circuiting in flocculation and sedimentation basins are discussed later in this chapter in the section on clarification.

Flocculation intensity (velocity gradient, *G*). Flocculation intensity determines both the rate at which particles come together to form a floc and the extent to which floc shear will occur and adversely affect the flocculation process. As the flocculation process proceeds, there is a progression toward larger particles that are more susceptible to shearing effects unless the flocculation intensity is gradually reduced.

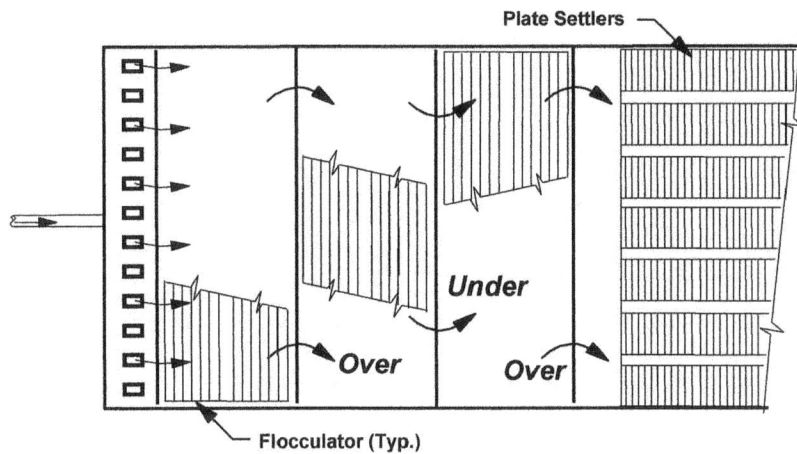

Source: Black & Veatch.

Figure 4-3 Crossflow baffling configuration for horizontal flocculators

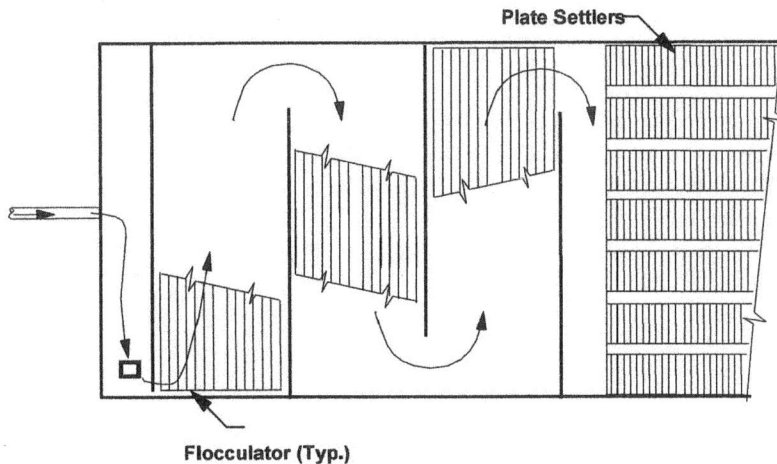

Source: Black & Veatch.

Figure 4-4 Top view of plug flow baffling configuration for horizontal flocculators

This is typically achieved through a staging of flocculation that allows a progression from higher to lower levels of flocculation intensity (energy input). In essence, the flocculation intensity is controlled at each stage to provide the best balance between the effect of agitation in promoting particle contact and the floc shear that becomes more significant as floc size is increased through successive stages. The term *tapered flocculation* is often used to describe this approach.

Adjustment of flocculation intensity for best conditions in each stage is an important consideration in both design and operation of flocculation facilities. This may even involve seasonal adjustments to accommodate changes in water quality and the associated changes in floc character. These types of adjustments are specifically practiced in some locations where a seasonal variation in cold water coagulant performance is encountered.

As indicated previously, maintaining constant G values does not ensure the same mixing quality as the scale of the process changes (Amirtharajah et al. 1991); therefore, flocculation intensity required in a jar test procedure can be different than that required in full-scale. Mixing intensities for proper jar testing results that simulate full-scale facilities can often vary from approximately 12 to 35 rpm, which corresponds to calculated jar test G values of approximately 3 to 30 s^{-1}, depending on the water temperature (Cornwell and Bishop 1983). General guidelines for full-scale flocculation basins are G values that vary between approximately 10 and 70 s^{-1} (Kawamura 2000). Often, in full-scale facilities where the level of short-circuiting is not well defined and basic information on the relation of process performance to G values has not been established, general guidelines for G values may be used for setting initial levels of flocculation, followed by minor adjustments until a best condition for operation is achieved. It should also be recognized that it is not uncommon for optimum flocculation conditions to vary seasonally, and examination of the value of seasonal adjustments may be appropriate at some locations.

Variation in water quality.   Water quality variations often exert a significant influence on flocculation requirements at a given location. In design, it is important that consideration be given to potential worst-case conditions. Cold-water coagulation and other factors such as algal events can influence flocculation efficiency in many plants, and it is important that the effect of these types of events be identified in selecting the appropriate flocculation conditions.

Note that during much of the year, flocculation performance may be less sensitive to facility configuration than during the most severe sets of conditions. Detention time, baffling, and staging to meet the needs of the more difficult treatment conditions often dictate the most appropriate facility configuration rather than a less rigorous configuration that produces what appears to be adequate performance under more typical conditions. The importance of seasonal considerations is illustrated in an example shown in Figure 4-5. This type of plot applies raw and settled turbidity data to give an indication of the effectiveness of a mixing/flocculation/clarification sequence for controlling turbidity. Where effective treatment conditions are encountered, a treatment plant is efficient in removing turbidity over a wide range of conditions. In essence, an efficient plant converts incoming particles to removable floc irrespective of the number of particles entering the treatment sequence, and the increase in clarified turbidity is limited as raw water turbidity is increased. Comparison of treatment efficiency that is indicated by this type of analysis can be seen in the results for three parallel treatment trains in Figure 4-5.

Trains 1 and 2 have similar overflow rates in sedimentation but have different flocculation configurations. The flocculation basin in train 1 has a design detention time of 32 min and is staged to provide a relatively good plug flow condition, while the flocculation basin in train 2 has a design detention time of 18 min and is not well

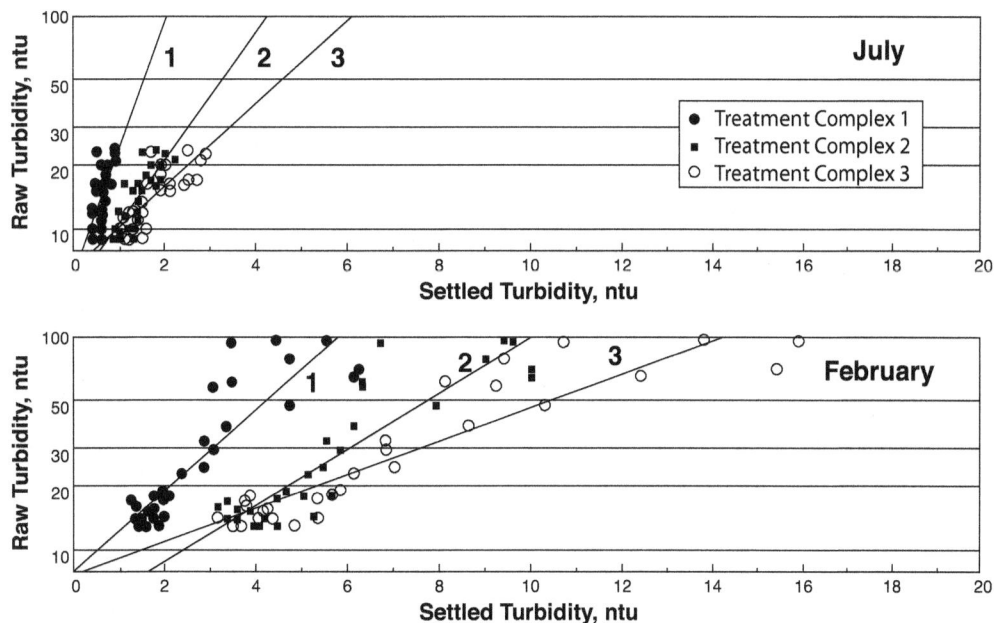

Courtesy of George Budd.

Figure 4-5   Seasonal variation in treatment for different flocculation configurations

baffled. The comparative effects are illustrated by the lower reduction in turbidity for any given raw water turbidity that is observed for train 2 in Figure 4-5. Train 3 has a poor flocculation configuration that is similar to train 2, but it has the additional limitation of a much higher loading on the sedimentation basins that further degrades treatment, as can also be seen in this figure. One particularly noteworthy aspect of this figure is the difference in seasonal response in comparing February and July results. These data are typical of many locations in that the colder months tend to be the more critical times for performance. In this case, flocculation performance during the summer was generally acceptable but became much more critical during the winter, when the need for improvements in trains 2 and 3 can more clearly be seen.

Preoxidant addition. Because of recent changes in disinfection by-product (DBP) regulations, many utilities that have traditionally added a preoxidant (in particular, free chlorine) have been forced to consider relocating their preoxidant feed point further downstream in the process. Utilities that make this type of change may experience some minor changes in their coagulation/flocculation process, including reduced settled water clarity, slightly increased coagulant doses, and slightly shorter filter runs. These changes are thought to be caused by the effect of the preoxidant on organic matter (as evidenced by the change in UV_{254} often observed after the addition of an oxidant), in particular the charge content of the organics. Minor adjustments in flocculation conditions can typically minimize the effects of this type of change in preoxidant addition location.

Physical observation of floc. Physical observation of floc can be an important part of flocculation process control, particularly during times of changing raw water quality. While the desired physical appearance of a floc can vary greatly depending on the raw water source, coagulant used, and the downstream clarification processes, one measure of good flocculation conditions is often the observation of "seams" or "marbling effect" during flocculation. This effect generally consists of defined clusters of floc with "clear" water evident between the clusters.

CLARIFICATION

Concepts of Sedimentation and Flotation

A concept of settling proposed by Alan Hazen about a century ago is still used to explain sedimentation of particles in basins. As particles pass through the settling basin in which water is flowing horizontally, those that settle to the bottom of the basin before leaving the quiescent zone are removed. A particle that enters at the water surface needs the greatest settling velocity to be removed before it reaches the end of the quiescent zone in the basin. According to Hazen's concept, particles would settle to the bottom and be removed more quickly in shallow basins. A very shallow basin could carry a large flow only if it was very wide, but such basins are not hydraulically stable, so the idea of using very shallow basins did not become practical until tube settlers and plate settlers were developed (Hansen and Culp 1967).

The simplifying assumptions of Hazen's concept are that water flows horizontally across the quiescent zone under plug flow conditions, each particle in the water is discrete, and no flocculation occurs during settling. These simplifications are not likely to be totally applicable in a real sedimentation basin. If flocculation does occur as particles settle and if the larger flocs have the same density as the smaller flocs before aggregation occurred, the larger flocs should settle faster than smaller ones. Thus, in a deep sedimentation basin, in contrast to a jar test jar, sedimentation velocity has the possibility of increasing with depth. Samples are extracted after only a short travel distance in the jar test procedure, so additional flocculation is much less likely in the test jar as compared to the full-scale basin. This may be offset, though, by nonhorizontal flow in a full-scale basin, whereas if the jar test jar is manipulated carefully, vertical currents can be minimized.

Summarized very briefly, the concept for flotation is similar to that for sedimentation. If particles can rise to the water surface or to the accumulated floc resting on the water surface and protruding slightly into the water before the water flows out of a flotation process basin, those particles will be retained in the basin to be removed later when the floated floc is removed.

Short-Circuiting in Basins

The ideal flow pattern for water in flocculation and clarification basins is plug flow, in which the detention time for all portions of water is the same from entry to exit of the basin. In actual practice, attaining perfect plug flow is not practical, but designers attempt to approximate this flow regime when designing basin inlet and outlet configurations and basin baffling. When plug flow is not attained, short-circuiting is said to occur. In this flow condition, some water passes through the basin in a time shorter than the theoretical detention time (basin volume divided by rate of flow), and other water remains in "dead zones" or zones of little circulation and flow for times that can be considerably longer than the theoretical detention time. Serious short-circuiting problems can impair process basin performance. Such problems can be continuous, as a result of basin design, or they can be intermittent, caused by environmental and water quality conditions. Strategies for dealing with continuous and intermittent short-circuiting are different.

If short-circuiting problems impair process performance on a continuing basis, an investigation of short-circuiting may be undertaken. A preliminary approach to this for sedimentation basins is to view the basin from a high point, looking down on the water. When water bodies are viewed from above, changes in color or turbidity are detected much more readily than when water is viewed from ground (water) level. Such high-level viewing might involve observing a basin from the roof of the highest building at the treatment plant, or possibly an overview from an aircraft. Photographs should be taken to provide documentation of the observations. Hudson (1981) showed two photographs (pages 73 and 94) that illustrated clouds of floc at inlets to settling basins. The pattern of the floc clouds can be an indicator of where unequal flow distribution is occurring. If basins are covered, visual observation of floc patterns would not be possible, and conducting tracer tests, as discussed in the following paragraph, could be more difficult because of restricted access for sampling.

When visual observation has indicated a potential short-circuiting problem, the extent of the problem can be documented by tracer studies as discussed by Kawamura (2000) in Appendix 9, "Tracer Test." Kawamura recommends using chemical tracers at concentrations that are safe from a public health perspective and that do not interact with any constituent in the water. One such tracer is lithium. Fluoride also can be used, but it can complex with aluminum when alum is used in high dosages, and lime softening has the potential to remove fluoride. If a tracer study reveals serious short-circuiting, the visual observations of floc clouds and tracer results may be good guidelines for making changes to inlet or outlet configurations. Tracer test procedures may need to be reviewed by drinking water regulatory agencies before such tests are performed. If treatment plant staff are unable to visually identify locations where short-circuiting is occurring, another option for addressing such problems is to obtain the services of a firm that can provide computational flow dynamics services and advice on how to remedy the short-circuiting.

Intermittent short-circuiting problems can be caused by sustained winds and by water quality changes. Just as a sustained high wind blowing for hours in one direction can cause turnover in a large lake, it also can do that in large, open settling basins, especially if the wind is blowing in a direction parallel to the long dimension of the basin. As with lakes, the longer the distance the wind can blow over the water, the greater the effect on flow patterns in the water. Wind-induced circulation in a sedimentation basin can upset the flow regime and induce short-circuiting.

Short-circuiting in sedimentation basins can be caused by density differences in water. For river sources, sudden high-turbidity episodes can result in a higher load of suspended solids and slightly greater density of water entering the basin. When this happens, the denser water can go to the bottom of the basin, displace water of lower density, and flow more rapidly to the basin outlet. Rapid changes in water temperature can be induced in lakes by sustained strong winds blowing onshore or offshore and causing turnover. If warmer water suddenly enters a basin, its lower density induces it to pass over the colder, denser water and reach the outlet sooner than normal. A change to colder, denser water would have the same effect as a density current caused by high concentrations of suspended solids, with the colder, denser water entering the basin, sinking to the bottom, and flowing to the outlet in a shorter time than normal.

Intermittent short-circuiting problems are difficult to deal with, but one possible remedy is to decrease the rate of flow in the plant, allowing more time for sedimentation to occur.

The Influence of Pretreatment Chemicals on Sludge Handling

The pretreatment chemicals used for coagulation or coagulants and polymers used in lime softening plants can have an effect on management of settled precipitates (sludge). A case study, "Palm Beach County Water Utilities Water Treatment Plant 8 Ferric Chloride Addition," describes how a metal coagulant was substituted for a polymer that had been used to assist clarification in a lime softening plant after difficulties were encountered with sludge handling. This study is presented in chapter 7.

Conventional Sedimentation Basins

Conventional sedimentation basins that provide several hours of detention time for clarification are being designed for new plants less often in an era in which a variety of high-rate processes are available, but in some instances economics still favors their application. Furthermore, many older plants have these basins. Detention times of 4 hr or greater were common at plants employing coagulation, whereas 2 hr of detention could be provided for lime softening. Some of these basins at plants using coagulation may not have sludge removal equipment, so the basins are removed from service and cleaned manually on a periodic basis. If sludge removal is performed manually, scheduling basin cleaning and maintenance during times when water demand is below average and raw water conditions are good is advisable in order to minimize the load on the remaining units in service. Sedimentation basins at lime softening plants are more likely to have sludge removal equipment because of the larger amount of sludge produced at these plants.

High-Rate Clarification Processes

Because of the area required for conventional sedimentation basins, design engineers and water treatment equipment suppliers developed concepts that could decrease the time required for sedimentation and thus decrease the size of the process basin needed. This was helpful at plants where land area was limited and also has been a useful approach when high-rate clarification processes are less expensive to build and operate than conventional basins.

Solids contact clarifiers and sludge blanket clarifiers. Early approaches to finding more economical alternatives to large sedimentation basins with long detention times led to the use of flocculation clarifiers and solids contact clarifiers. These high-rate processes combined a central mixing zone, flocculation, and an outer sedimentation zone in one basin. In some of these clarifiers a baffle wall separates the mixing and flocculating zones from the sedimentation zone. Chemical reactions and floc formation are promoted by recirculation of water within the mixing zone and by

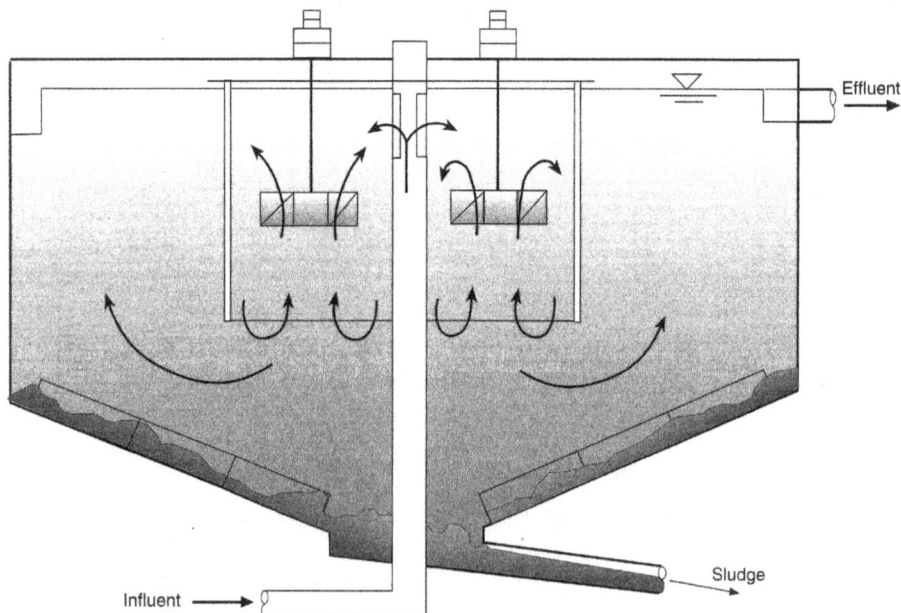

Source: WesTech Engineering Inc.

Figure 4-6   Flocculator/clarifier

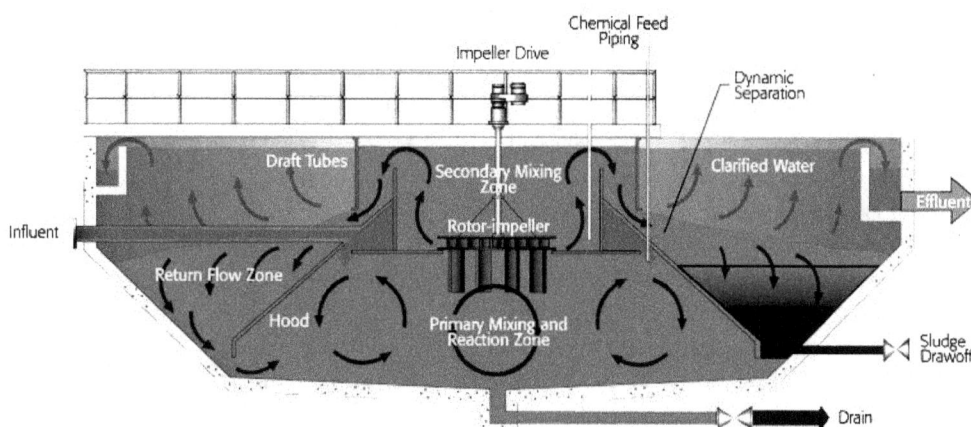

Source: Degremont Technologies.

Figure 4-7   Blanket clarifier

mixing incoming water with sludge drawn from the bottom of the basin. The units can achieve high solids in the mixing zone and may be operated with a sludge blanket in the settling zone. Figure 4-6 shows a flocculator/clarifier, and a diagram of a blanket clarifier is presented in Figure 4-7.

In sludge blanket clarifiers, floc that settles in the sedimentation zone is allowed to accumulate in the bottom of the clarifier. The level of this accumulation is controlled by periodic removal of excess floc from the clarifier. The accumulated floc is termed the "sludge blanket." As water from the flocculation zone enters the bottom of the sedimentation zone, it is forced to pass through the sludge blanket, which results in further contact between the floc leaving the flocculation zone and the floc in the sludge blanket, improving agglomeration of the floc and its separation from the water. The upward flow of water through the sludge blanket in the sedimentation zone causes the sludge blanket to expand. Some sludge blanket and solids contact clarifiers are designed so the cross section of the sedimentation zone increases as water moves upward toward the weirs or launders. This causes the rise rate to decrease and is helpful in preventing the blanket from washing out of the sedimentation zone when the rate of flow through the clarifier is increased. Another variation of the blanket clarifier process is pulsed flow, in which inflow to the clarifier basin takes place in periodic pulses, which are intended to impart an agitation to the floc blanket and enhance floc removal. This type of clarifier is shown in Figure 6-2 in chapter 6 of this manual. Blanket clarifier process units have been operated at overflow rates of about 1.8 gpm/ft^2 (4.4 m/hr) for softening and about 1.0 gpm/ft^2 (2.4 m/hr) for chemical coagulation applications. These overflow rates can be approximately doubled if tube settlers or plate settlers are added to the sedimentation zone. Further information on tube and plate settlers is provided below.

Blanket clarifiers and reactor clarifiers are intended to operate on a continuous basis, 24 hr per day, and if used at coagulation plants, are most effective if they can be used in a base-loaded mode. When this process equipment is operated at a constant rate of flow, the blanket level within the sedimentation zone should not rise or drop substantially. If the rate of flow decreases, the rise rate decreases, causing the blanket level to drop. If the flow increases, the higher rise rate will cause the top of the blanket to move closer to the effluent weirs or launders. Daily increases and decreases in flow in a blanket clarifier and, worse, on-and-off operation of such process equipment can cause serious problems, especially when floc in the blanket is only slightly more dense than water. Performance of blanket clarifiers and reactor clarifiers also can be impaired by rapid changes in water temperature, as was explained previously in the section on short-circuiting.

Tube settlers and plate settlers. Inclined tube settlers were developed in the 1960s as a practical approach to the problem of providing short distances for sedimentation while also providing for narrow sedimentation basins that had lateral hydraulic stability (Hansen and Culp 1967). They resolved the question of how to apply Hazen's sedimentation theory and increase the settling surface area to achieve settling in a short time. The tube settler concept involved placing a bundle of inclined tubes in a settling basin, positioned so that flocculated water passed upward through the tubes. The vertical distance inside the tube from the top to the bottom of the inclined tube (the distance a particle must settle to reach the "bottom" of the "basin") can be several inches (centimeters) instead of 10 to 16 ft (3 to 5 m) from the water surface to the bottom of a conventional settling basin. For example, a common angle of inclination for a tube is 60°, and for such a tube with an internal perpendicular distance of 2 in. (5 cm) from the top wall across to the bottom wall, the vertical distance from the top of the inclined tube's wall to the bottom of the tube's wall is 4 in. (10 cm). A bundle of tubes provides many false floors onto which floc particles can settle, agglomerate, and slough down to the bottom end of the tube, to be discharged and settle to the bottom of the settling basin.

Tube settlers were initially used in small preengineered package plants to maximize treatment capacity in a small volume. They have also been used to uprate conventional sedimentation basins as well as for new sedimentation basins specifically designed for the shorter detention time appropriate for basins with tube settlers. A diagram depicting tube settlers in a basin is shown in Figure 4-8. Gross overflow rates range from 1 to 3 gpm/ft^2 (2.4 to 7.3 m/hr), depending on the settling characteristics of the floc that is formed in pretreatment. Tube settlers are frequently added or retrofitted into solids contact clarifiers.

Tube settlers in small package or preengineered plants tend to be placed at shallow angles of inclination. They are cleaned by backflushing water through the tube settlers. Periodic flushing of the steeply inclined tube settlers may be needed if floc tenaciously adheres to the tubes and does not slough out.

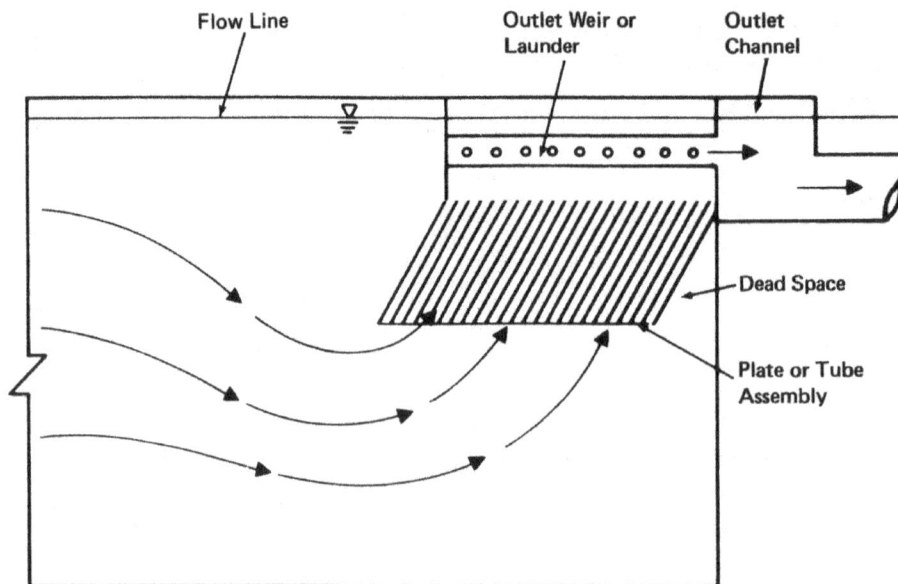

Courtesy of Kerry E. Dissinger, Brentwood Industries Inc.

Figure 4-8   Tube settler installed in sedimentation basin

Often tube settlers are fabricated from lightweight noncorroding material such as plastic. This eliminates corrosion problems and decreases the weight of the tube bundles that are supported near the upper level of the basin. If tube settler modules are made of plastic, which can be flammable, repair work that generates intense heat (e.g., welding) has the potential to deform tube settlers or cause them to burn. Their large surface-to-volume ratio can result in an intense, rapidly spreading fire.

Somewhat analogous to the inclined tube settlers is the European concept of inclined plate settlers, which operate with a gross overflow rates in the range of 2 to 4 gpm/ft^2 (5 to 10 m/hr). Treatment is enhanced by the formation of uniform-sized floc. Plate settlers have been used in large plants as well as in some package plants. These are not as easy to retrofit into existing sedimentation basins. The plates are typically about 7.5 ft (2.3 m) deep, so they require basins that are deeper than some conventional basins.

Contact clarifiers.   Contact adsorption clarifiers (CACs) have been used in some preengineered (package) plants, with a multimedia filter placed in series after the CAC unit, which consists of a bed of coarse filter material through which coagulated water passes. The flow of coagulated water through the coarse media bed aids in flocculation and removal of floc and prepares the water for filtration in a dual or mixed media bed. These clarifiers have been produced for operation in both pressure filtration and gravity filtration modes. Large CACs can serve as stand-alone pretreatment facilities ahead of conventionally constructed filters. Overflow (or filtration) rates for CACs and roughing filters can be as high as 10 gpm/ft^2 (24 m/hr), with application generally limited to low-turbidity sources with low coagulant demands. Attaining correct coagulation chemistry is as important for CAC units as it is for other clarification processes.

When used in smaller preengineered package plants, a CAC process unit is placed ahead of a multimedia filter, and the two processes are operated in series. This is shown in Figure 4-9. Most clarification processes (except for conventional settling basins requiring manual sludge removal) are operated continuously, and the settled sludge or floated floc is removed while the clarifier remains in operation. However, a CAC performs a filtration function, so a CAC requires periodic flow interruption to clean out accumulated floc. When head loss builds up faster in a CAC unit than in the granular media filter

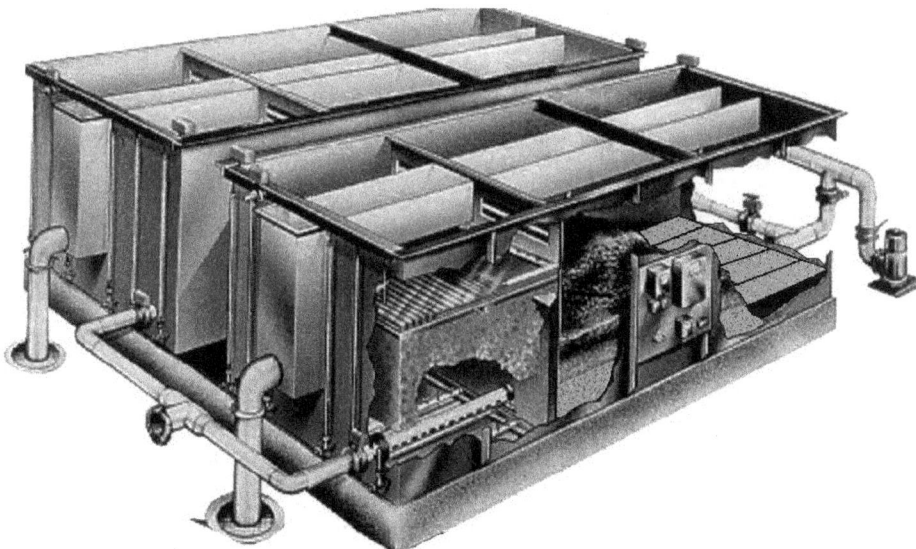

Source: Siemens Water Technologies Corporation.

Figure 4-9   Contact adsorption clarifier and multimedia gravity filter in series

following the CAC, the flow of clarified water into the filter must be interrupted while the CAC is flushed to clean out accumulated floc if the CAC and the filter are piped in series. This can result in shutting off the filter when the clarifier is cleaned. For a small system operating a single CAC unit and filter, if the filter is not backwashed each time the CAC unit is taken out of service and cleaned, operators should be very careful when restarting the partially clogged filter upon resumption of treatment.

Ballasted flocculation.   Ballasted flocculation is a process that applies a relatively high concentration of small-diameter sand to accelerate settling of floc formed by coagulant. Sand concentrations are typically in the range of 3 or 4 g/L up to 6 g/L. The process begins with an initial addition of coagulant to condition incoming particles and lead to the formation of small masses of floc that are then attached to the sand using a polymer. The first physical step in the sequence includes dispersion of the coagulant along with the addition of pH-adjusting chemicals if required. The second step includes mixing for 1 to 2 min to produce a small floc. The third step is an additional mixing step of 1 to 2 min in which sand is added, and contact between the sand and floc begins. Polymer is then applied so that floc attachment to the sand can take place. Optimum location for polymer addition can vary, and it is common to provide flexibility for several points of addition. A diagram of a ballasted flocculation process train is shown in Figure 6-3 in chapter 6 of this manual.

Following the floc formation and sand addition steps, a mixing step of 3 to 6 min is applied to build floc size. The high density suspension that results is conveyed to the settling tank where solids liquid separation takes place at loading rates that can be in the range of 20 gpm/ft^2 (49 m/hr) or greater. The commercial version of this process applies tube settling modules within this unit to enhance the settling process.

A critical aspect of the process is collection of the sand/floc mass at the bottom of the settling unit for recycle to separation devices that will recover sand from this stream, while discharging less dense floc that is wasted from the system. In this manner, sand loss from the system is minimized. Hydrocyclones have been applied for this purpose in the commercial version of this process. Sand recovery using this approach can be relatively effective, requiring limited addition of about 1 mg/L or less to make up for overall losses across the system.

Short detention time in the mixing units combined with high overflow rates produce a clarification sequence that has a very small footprint. The commercial process has also been subject to recent innovations that are intended to improve hydrodynamic features, improve overall process dynamics, and further reduce process footprint. This sequence combines the sand injection and final mix step, thereby eliminating one of the two initial mix tanks.

Application of proper coagulant dose and pH is a critical underpinning for this process, as is the case for all clarification processes. Selection of an appropriate polymer (Van Cappellan et al. 2008), point or points of polymer application within the sequence, and polymer dose are also essential. Setting polymer dose requires understanding of certain basic features of operation. Dose that is too small will not adequately attach floc to the sand, while excess polymer can carry over to downstream filters and potentially have adverse effect on filter run times. Selection of sand size can also be important (Van Cappellan et al. 2008). Typical sizes for water treatment are less than those for wastewater applications, and care should be exercised to avoid sand of excess size.

Dissolved air flotation.   The DAF process is usually considered when the raw water is characterized by contaminants that are or produce low-density solids. This would include solids that are created by forming hydroxide flocs when feeding inorganic chemicals for reducing high color or organics and the solids created by oxidizing soluble iron and/or manganese. Filter backwash water contains high-turbidity but low-density floc that is readily clarified in the DAF process. In addition, the DAF process is preferred for algae reduction since it takes advantage of algae's natural buoyancy to remove it. Gregory et al. (1999) reported that removal of the algae *Microcystis* was 98 percent, as compared to 76 percent for sedimentation. The presence of heavier particles such as heavy silt from runoff tends to reduce efficacy of DAF, as high concentrations of turbidity-causing mineral particles are difficult to float when coagulated. This process has been used to treat lowland mineral-bearing river water with turbidity up to 100 ntu (Gregory et al. 1999).

The DAF process starts by feeding coagulant to the raw water by either a hydraulic (baffles or in-line mixer) or mechanical rapid-mix chamber designed to achieve rapid dispersion of the coagulant. This is followed by flocculation in one to three stages, with two-stage flocculation being the most common. Total flocculation hydraulic retention times generally vary between 5 to 20 min of design flow, with tapered velocity gradients from 50 to 100 sec^{-1} in the first flocculator to 20 to 50 sec^{-1} in the last flocculator. Unlike conventional sedimentation, where larger, more settleable floc is desired, the goal of the flocculation process ahead of DAF is to form a smaller pin floc that will readily attach to the bubbles that are introduced in the DAF process.

After a floc is formed, the flocculated water stream is injected with water that has been saturated with air. In the DAF process, a clarified-water sidestream of 5 to 15 percent of the raw water flow is pumped into a saturator, where it is mixed with high-pressure air (80 psi or 560 kPa) and then stored in the bottom of the saturator tank. From the bottom of the saturator tank, water supersaturated with air can be returned to the dissolved air flotation clarifier reaction zone, where the pressure is dropped to atmospheric through specially designed orifices or nozzles in a distribution header that spans the width of the tank to generate bubbles in the size range of about 10 to 100 μm to float flocs (Zabel 1985). The microbubbles combine with the floc and carry them to the surface of the flotation cell. In a properly operated unit, the water containing the microscopic bubbles looks milky. Large bubbles and turbulent, frothy water are indicative of failure to form the microscopic bubbles that carry floc to the water surface and are a symptom of serious process problems likely related to inadequate air pressure in the saturators. While conventional DAF systems were typically designed at loading rates near 4 gpm/ft² (10 m/hr), depending on the raw water to be treated and the design of the DAF, the flotation loading rates of current high-rate DAF systems can typically vary from 4 gpm/ft² (10 m/hr) to 12 gpm/ft² (29 m/hr), with rates up to 16 to 20 gpm/ft² (39 to 49 m/hr) under certain conditions. A diagram of a DAF process train is shown in Figure 4-10.

Unlike blanket clarifiers, DAF clarifiers do not require steady-state operation. However, if flow through the process is changed, the recycle rate of flow may need to change as well, to maintain the same percentage of recycle. If changes to the recycle rate are made, care must be taken to ensure that flow through the nozzles is within the nozzle operating range. Often multiple recycle headers are constructed to provide flexibility in adjusting recycle rates to ensure optimal operation of the nozzles and subsequent formation of small bubbles. Bubbles that are too large have a greater tendency to disturb floated floc and prevent floc from rising to the top of the water.

Another operating parameter that affects process performance is saturator pressure. A sufficient volume of dissolved air should be provided to form air bubbles capable of lifting the suspended solids in the flocculated water to ensure successful operation of the clarifier. This is attained by a combination of the percentage of recycle flow through the saturator and the air pressure maintained in the saturator. Higher concentrations of suspended solids, such as algae or turbidity in the raw water, may need a greater volume of dissolved air for effective removal. In addition, low levels of dissolved oxygen in the raw water can create the need for higher percentages of recycle; in these cases,

5–20 min floc time   **8–20 gpm/ft^2 loading rate**

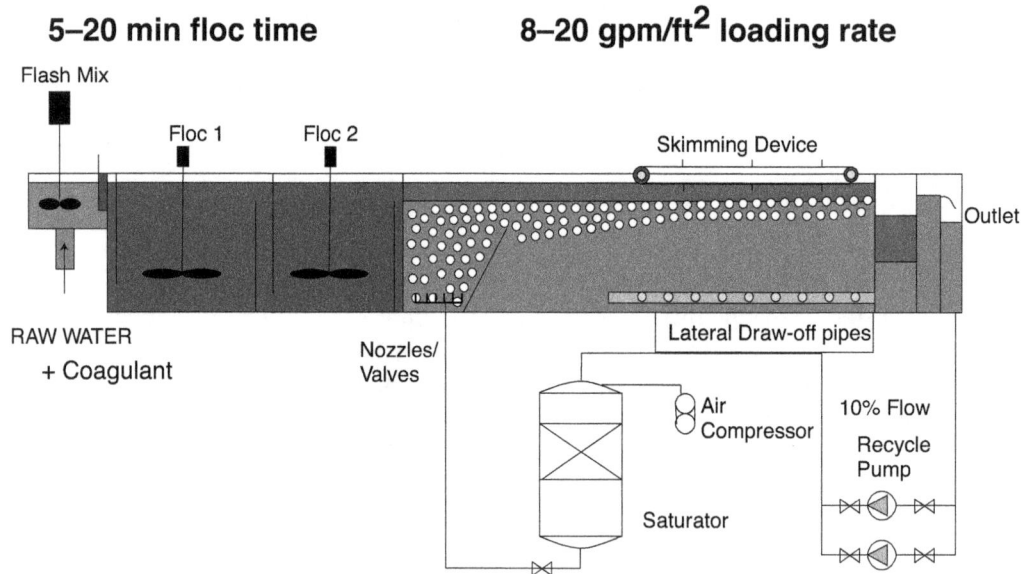

Compressed air dissolved in water at 80 psi

Source: ITT Water & Wastewater.

Figure 4-10   DAF process train

it is generally more cost-effective to utilize another method of increasing raw-water dissolved-oxygen concentration, such as reservoir aeration or some type of in-line aeration, rather than use the DAF recycle system to satisfy oxygen deficit.

Correct management of pretreatment chemistry is a must for effective operation of a DAF clarifier, and as discussed in chapter 2, chemical dosages can be determined by flotation jar tests. In a pilot plant evaluation of both direct filtration and dissolved air flotation followed by filtration (Black and Veatch 1994), the pretreatment chemistry that was working well for direct filtration was used when the DAF clarifier was brought online for testing, and very effective treatment was attained in the DAF unit. In fact, within a half hour of startup, the clarified turbidity produced by the DAF process was 0.2 ntu. This demonstrated the ability of this process to be brought online quickly and achieve effective treatment if correct pretreatment chemistry has been identified.

As a DAF clarifier is operated, floc is carried to the top of the water by the air bubbles, where the floc accumulates. A layer of scum forms on the water surface and gradually grows thicker as the operation continues. The blanket of sludge is supported from beneath by tiny air bubbles, but at some point in time the accumulated floc, called *float*, becomes so thick that it must be removed from the water surface before it begins to break apart and sink back down into the clarified water. As time passes, the float next to the clarifier wall may begin to adhere to the wall. If this happens, spraying water on the clarifier wall for a short time before the float is removed and in the early stage of removal can help alleviate this condition. The float is removed periodically by either a mechanical scrapper or by hydraulic means. While more complex, the mechanical sludge removal system usually results in obtaining sludge of 2 to 5 percent dry weight basis. This high solids content reduces the quantity of sludge to be processed in the residuals handling portion of the plant; however, some water may have to be added back to the thicker sludge attained with mechanical sludge removal to allow it

to flow into pumping units and be pumped to subsequent processes. The clarified effluent water is drawn off the bottom of the tank.

Maintenance tasks at a DAF plant involve recycle pumps, air compressors that
maintain pressure in the saturator, the condition of the saturator, the nozzles that
deliver supersaturated water into the clarifier, and the mechanical sludge removal
system. Periodic inspection of the saturator packing to verify that there is no buildup
of precipitates and slimes should be performed.

DAF units may also be fitted directly above mono-, dual-, or tri-media filters. These
units are usually limited in the loading rate to the acceptable loading rate of the granular media filters (Kawamura 2000). The flotation cell *must* be deep enough to contain
the backwash troughs, 30 to 36 in. (0.76 to 0.91 m) of media, an allowance for media
expansion during backwash, and the underdrain system required for filtration, all of
which would be located beneath the separation zone where clarification by flotation is
achieved. In addition, the control panels would need to handle both processes. When
a filter backwash is required for process equipment that provides DAF clarification in
the same basin as the filter, the flow of pretreated water into the process basin must
be stopped, and if this involves interruption of raw water flow, dosing of coagulant also
must be stopped during the backwash. Furthermore, before the filter is backwashed,
the floated floc (sludge) needs to be removed for disposal so it is not excessively diluted
by being combined with the washwater. Figure 4-11 shows a treatment train in which
the DAF clarification step is accomplished above the filter bed.

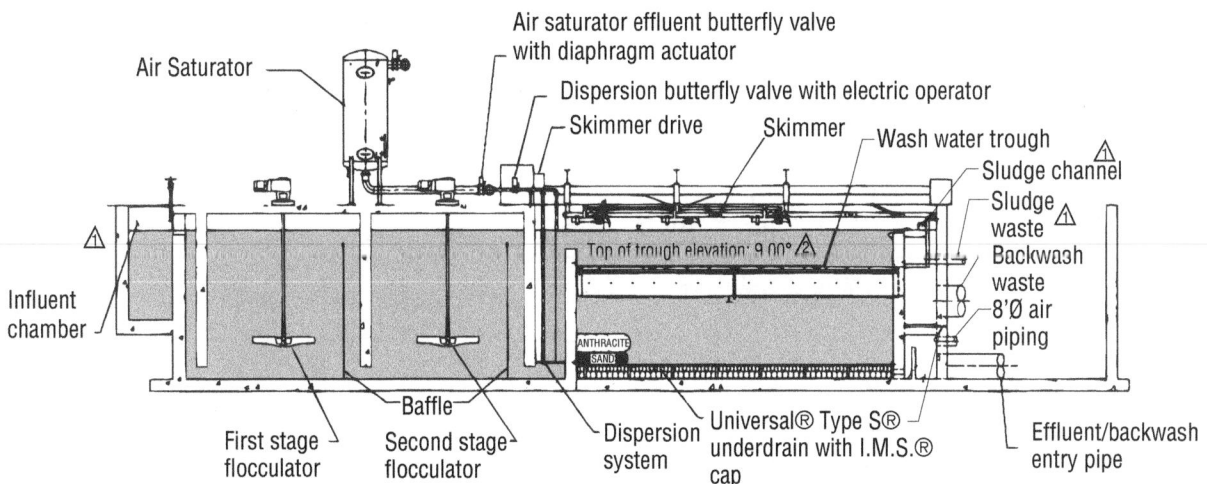

NOTES:
1. All field wiring, conduit, supports, hardware
   and process supply piping not by Leopold.
2. All walls shown as 12' thick unless otherwise noted.

Source: ITT Water & Wastewater.

Figure 4-11   In-filter DAF treatment train

REFERENCES

Amirtharajah, A., M. Clark, and R. Trussell. 1991. *Mixing in Coagulation and Flocculation.* Denver, Colo.: AWWA and Awwa Research Foundation.

Black and Veatch. 1994. Preliminary Engineering Report—Phase II: Pilot Plant Testing for Filtration of Table Rock and North Saluda Reservoirs, p. VIII–3. Submitted to Greenville Water System, Greenville, S.C.

Clark, M.M., R.M. Srivastava, J.S. Land, L.J. McCollum, D. Bailey, J.D. Christie, and G. Stolarik. 1994. *Selection and Design of Mixing Processes for Coagulation.* Denver, Colo.: AWWA and Awwa Research Foundation.

Cornwell, D.A., and M.M. Bishop. 1983. Determining Velocity Gradients in Laboratory and Full-Scale Systems. *Jour. AWWA,* 75(9):470–475.

Foellmi, S., and H. Bryant. 1991. Lamella Plate Settlers, Design and Operation—Two Case Histories. In *Proceedings of 1991 AWWA Conference.* Denver, Colo.: AWWA.

Gregory, R., T. Zabel, and J. Edzwald. 1999. Sedimentation and Flotation. In *Water Quality and Treatment,* ed.

R.D. Letterman. 5th ed. (7.1–7.87). New York: McGraw-Hill.

Hansen, S.P., and G.L. Culp. 1967. Applying Shallow Depth Sedimentation Theory. *Jour. AWWA,* 59(9):1134–1148.

Hudson, H.E., Jr. 1981. *Water Clarification Processes: Practical Design and Evaluation.* New York: Van Nostrand Reinhold.

Janssens, J. 1992. Developments in Coagulation, Flocculation, and Dissolved Air Flotation. *Water / Engineering and Management,* 139(1): 26–31.

Kawamura, S. 2000. *Integrated Design and Operation of Water Treatment Facilities.* 2nd ed. New York: John Wiley & Sons.

Van Cappellan, J., J.K. Clement, S. Kempeneers, G. Budd, W. Bossaerts, M. Huysmans, and G. Milton. 2008. Long Term Comparison Testing with Sand Ballasted Flocculation (SBF) and High-Rate Dissolved Air Flotation (DAF). *IWA Water Practice and Technology,* 3(4).

Zabel, T. 1985. The Advantages of Dissolved-Air Flotation for Water Treatment. *Jour. AWWA,* 77(5):42–46.

Chapter **5**

Filtration

Kevin Castro and Rasheed Ahmad

INTRODUCTION

Filtration is a solid/liquid separation process utilized for water treatment. The mechanisms for filtration of solids (particles) are complex and are influenced by the type of filter used. Granular media filters (depth filtration) primarily remove coagulated particles through sedimentation, interception, and diffusion. Granular activated carbon (GAC) filters have similar removal characteristics for particles and can also remove dissolved species through adsorption. Slow sand and biologically active filters provide biodegradation of dissolved species in addition to depth filtration. Membrane filters provide a mechanical sieving of particles smaller than the pore size of the membrane. Cake filtration (e.g., diatomaceous earth filtration) provides deposition of particles on a precoated filter medium, where the deposited particles become part of the filter medium.

The range of sizes of uncoagulated particles that can be removed varies. Figure 5-1 summarizes the general size range of different types of granular media filters and pore size for other filtering materials.

Rapid-rate granular media filters using sand, anthracite, or GAC are capable of removing particles that are much smaller than 100 µm, if the particles have been properly coagulatedf prior to filtration. Polioviruses are about 25 nm in size (Young 1961), yet research by Robeck et al. (1962) demonstrated that poliovirus removal by alum coagulation and filtration could range from 90 to 99 percent, whereas without coagulation removal was erratic, ranging from 1 to 50 percent. With conventional treatment (coagulation, sedimentation, and filtration) removal exceeded 99.7 percent. Research on asbestos fiber removal at the Duluth, Minn., water filtration plant (Logsdon 1979), showed that submicron amosite particle removal by conventional treatment could be as high as 99.99 percent. Likewise, removal of particles larger than 1 µm by coagulation, flocculation, and rapid-rate filtration could exceed 99.9 percent for *Giardia* cysts (Logsdon et al. 1981) and for *Cryptosporidium* oocysts (Patania et al. 1995). These reported results contrast greatly with the uncoagulated particle removal capabilities

Particle Diameter

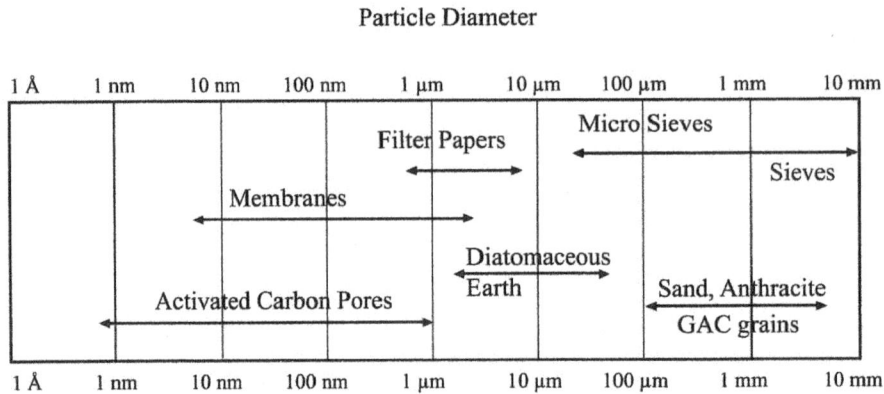

Courtesy of Kevin Castro.

Figure 5-1   Size Range of Various Filter Types

presented in Figure 5-1 and thus dramatically illustrate the importance of attaining effective coagulation when managing rapid-rate granular media filtration plants. The role of coagulation at such plants is so crucial that the Surface Water Treatment Rule (SWTR) does not allow any removal credit for *Giardia* cysts at a rapid-rate filtration plant that does not practice coagulation prior to filtration.

PRETREATMENT

The type of filtration utilized and the quality of the raw water influence the extent of pretreatment required for effective filtration of surface waters containing higher levels of solids or groundwaters with high metal concentrations. Filtration typically follows a coagulation, flocculation, and sedimentation process. An oxidation/precipitation process is employed when dissolved metals are present. For higher-quality waters, limited or no pretreatment is necessary to provide effective filtration (primarily membranes, cake filtration, and slow sand filtration). Filtration rate and effective size of media for the filter can also influence the extent of pretreatment necessary. Higher-rate filters will generally require better raw water quality or improved pretreatment to reduce loading on the filters.

The primary goal of pretreatment processes ahead of filtration is to remove sufficient particulate matter before the filters to extend filter run length (time between filter backwashes); and/or to provide oxidation, destabilization, and/or flocculation of particles and organics to achieve a size that is filterable by the medium.

For granular media rapid-rate filters (the primary focus of this chapter), the pretreatment approach will generally consist of coagulation, flocculation, and clarification. For lower-turbidity waters or waters containing algae, dissolved air flotation (DAF) may be used as the clarification process. Some type of sedimentation generally is used for higher-turbidity waters. The pretreatment for granular media rapid-rate filters should provide sufficient particle removal or particle flocculation ahead of the filters to provide reasonable filter run lengths (>24 hr under peak flows). Another criterion for filter performance is the unit filter run volume (UFRV), which is the volume of water filtered through a unit surface area during the run. UFRV goals may range from 5,000 gal/ft^2 to 10,000 gal/ft^2 or higher (200 m^3/m^2 to 400 m^3/m^2 or higher). A generally accepted pretreatment performance goal for many granular media filtration plants is to achieve a filter influent turbidity of less than 1 ntu 90 percent of the time and an effluent turbidity less than 0.1 ntu.

PARTICLE REMOVAL IN RAPID FILTRATION_____

The mechanisms of particle removal through a rapid-rate granular media filter are dependent on particle size. Particles larger than the void spaces in media are removed through physical straining. Particles smaller than the pore size of media are removed by sedimentation, interception, and diffusion. Figure 5-2 depicts mechanisms of particle capture on a spherical media grain.

Sedimentation

Particles with density greater than that of water tend to react primarily to gravitational forces. The ratio of settling velocity to superficial velocity has been shown to define the removal efficiency. The importance of sedimentation in filtration diminishes with increased velocity through the filter.

Interception

Interception occurs as finite-sized particles are transported to the media surface as the fluid streamlines constrict. Interception increases as the ratio of particle size to media size increases.

Diffusion

Brownian motion will cause a smaller particle to deviate from its streamline to the media surface. Diffusion is the most dominant mechanism for particle removal for particles <1 μm.

Destabilization (charge neutralization) of particles in the coagulation step assists in attachment of particles to the media, as explained in chapter 1.

The collection efficiency as derived from Rajagopolan and Tien (1976) as a function of particle diameter for varying superficial velocity for a 0.5-mm spherical collector (media) is shown in Figure 5-3.

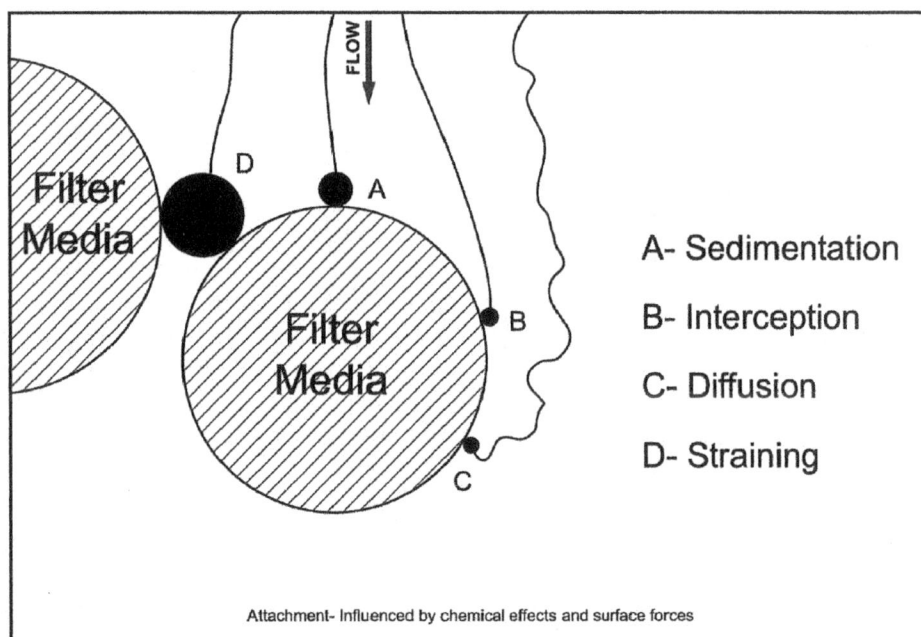

FLOW

Filter Media

Filter Media

D

A

B

C

A- Sedimentation

B- Interception

C- Diffusion

D- Straining

Attachment- Influenced by chemical effects and surface forces

Courtesy of Kevin Castro.

Figure 5-2   Particle removal in a granular filter

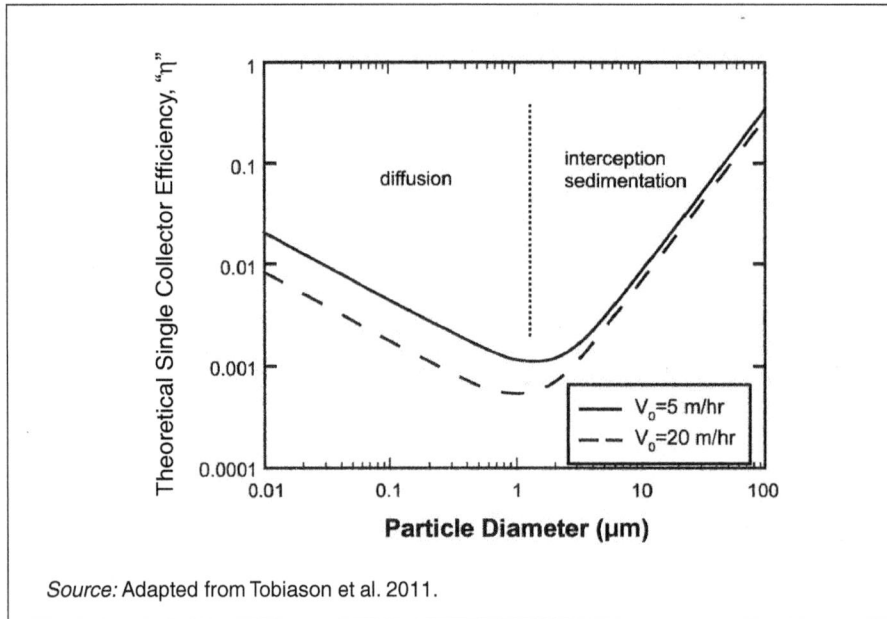

Source: Adapted from Tobiason et al. 2011.

Figure 5-3   Calculated clean-bed single-collector removal efficiency as a function of particle size for two filtration velocities according to Tufenjki and Elimelech 2004 (d_c = 0.5 mm, V = 5 and 20 m/hr, T = 20°C, ρ_p = 1,050 kg/m³, ε_0 = 0.4, H = 1.0 × 10^{-20}J)

An important conclusion from the relationship shown in Figure 5-3 is that minimum particle removal efficiency occurs at particle sizes of 1 to 2 µm. Figure 5-3 also indicates reduced particle removal efficiency at higher filtration rates, particularly for smaller particles.

Biological Filtration

Biological filtration for drinking water treatment has become more common because of increased use of ozone as disinfectant. The increased use also stems from recent stringent water quality regulations. In a conventional water treatment plant when ozone is used as a predisinfectant, the rapid gravity filters will become biologically active. Without using ozone, there are a variety of approaches to achieve biological filtration or biological treatment of drinking water. Biological treatment can be achieved through slow sand filtration, bank filtration, ground passage, and granular activated carbon contact. Although the approaches differ substantially in terms of process configuration, they all have two common and critically important characteristics. The first characteristic is that all of the processes are of biofilm type in which bacteria are stably retained in the process by natural attachment to a solid surface such as sand, anthracite, or granular activated carbon. The second characteristic is that the processes select for oligotrophic bacteria or those bacteria especially adapted to function when their substrate (food) concentrations are very low. Oligotrophy is important because organic substrates are present in drinking water supplies at microgram per liter levels. Most of the biological processes are aerobic, which means that dissolved oxygen is present and utilized as the electron acceptor by the bacteria. For this reason, it is important to keep biological filters in operation continuously to the extent practical, and they should not be out of service for a period of time long enough for the dissolved oxygen in the water within the filter bed to be depleted, as this would be harmful to the biological population and could inhibit biological removal of contaminants for a while after resumption of filtration (Ahmad et al. 1998, Bouwer and Crowe 1988, Hozalski et al. 1995, Mallevialle et al. 1992, Rittmann and Snoeyink 1984, Urfer et al. 1997).

Benefits of biological filtration include decrease of the potential for bacterial regrowth in the water distribution system, reduction of chlorinated disinfection by-products (DBPs), reduction in chlorine demand, and decrease of corrosion potential. So, biological filtration is used to achieve three broad goals: (1) biologically oxidize biodegradable components, making the water biologically stable and reducing the need for excess chlorination, which will in turn reduce the formation of DBP compounds that are suspected or potential human carcinogens; (2) biodegrade synthetic organic micropollutants that are harmful to human health; and (3) remove nitrate and nitrite via denitrification. In most drinking water treatment, however, the first two goals are main objectives of biological filtration. Particle (silt, clay, precipitates) removal does occur in biological filters even though it may not be an intentional goal, and biological filters have been shown to be effective for removal of particulate matter, including pathogens. Emelko et al. (2006) reported that in a full-scale evaluation of biological filtration, the biological filters, whether GAC filter adsorbers or dual media filters, consistently met high standards of particle removal as indicated by production of filtered water turbidity less than 0.1 ntu regardless of water temperature or backwash protocol employed. Amburgey et al. (2005) tested pilot-scale filters for removal of 4.5-μm polystyrene microspheres and *Cryptosporidium parvum* oocysts and found that biological filters have a slight advantage over conventional filters in removing pathogens such as *Cryptosporidium*.

It is well established that ozonation increases the fraction of natural organic material (NOM) that is biodegradable. Thus, the increase in biodegradable organic matter (BOM) upon ozonation generally considerably enhances biological activity in filters following ozonation. Often, biological filtration can reduce BOM concentration to approximately preozonation levels, although this depends on the specifics of biological filters and water quality parameters, and the composition of BOM may be different after biological filtration. Currently identified ozonation by-products such as aldehydes, carboxylic acids, and keto acids are biodegraded by biological filters with over 75 percent removal efficiency. Several synthetic organic compounds (SOCs) are also substantially biodegraded by the biological filtration process. In particular, phenol, chlorinated phenols, and chlorinated benzene show significant percent removals immediately or after a short acclimation period (Manem and Rittmann 1992).

FILTER OPERATION AND MANAGEMENT

Coagulation and pretreatment processes prepare water for filtration. The final aspect of process management related to coagulation and filtration is management and control of the filter when it is placed into service after backwashing, during the filter run, and when the filter is backwashed at the end of the run. The benefits of performing coagulation and pretreatment carefully and correctly can be lost if a filter is not controlled and managed properly. This section presents information on managing filters for the purpose of attaining excellent filtered water quality.

Returning Filters to Service After Backwash

The tendency of filters to produce water with higher-than-normal turbidity upon being returned to service has been documented (Amirtharajah and Wetstein 1980) for nearly three decades. Deposited particles become collectors or part of the filter media after they have become attached to media grains or to previously attached particles. Filter backwashing removes captured particles that can assist in additional particle removal.

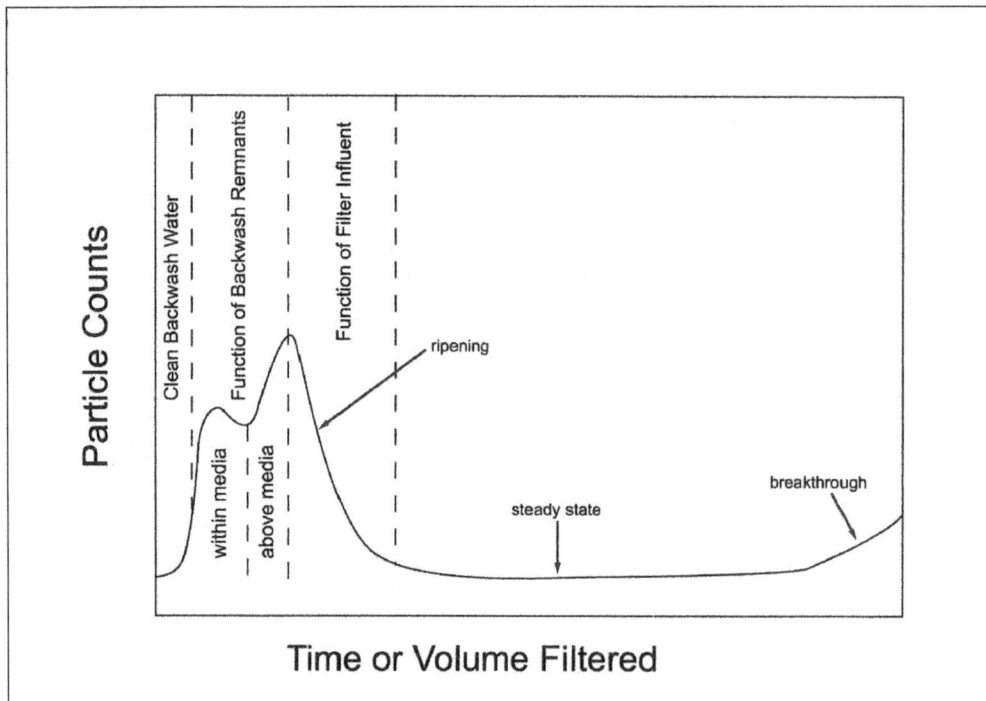

Source: Adapted from Amirtharajah and Wetstein 1980.

Figure 5-4   Typical filter breakthrough curve

In addition, remnants of floc dislodged from filter media during filter washing, but not washed out of the filter box, can be discharged in filter effluent. Amburgey et al. (2003) have presented experimental data showing that floc remnants can become more electronegative as backwashing progresses and thus become less likely to reattach to media grains when filtration resumes after backwashing. Figure 5-4 from Amirtharajah and Wetstein depicts a generalized particle breakthrough pattern over time after a backwashed filter is returned to service. Note that the time scale is not uniform but is expanded to show details of filter behavior when the filter is returned to service and later is condensed to present hypothetical trends during the complete run and beyond.

With standards on individual filter effluent turbidity in the operation of granular media filters, it is important to recognize the potential for particle breakthrough and to exercise appropriate operational procedures and costs to minimize it. There are several methods of minimizing particle breakthrough after a filter backwash, including: (1) modified backwash techniques, including adding chemical to backwash water and modifying wash-water rise rates and backwash duration; (2) adding chemical to filter influent as the filter box refills following backwash; (3) filter resting; (4) filter-to-waste; and (5) starting a filter at a low rate and gradually increasing the filtration rate.

At some water utilities, a single approach of those listed above is used to help control the initial turbidity spike when a filter is returned to service after washing. Others use a combination of two or three techniques to enhance production of low-turbidity water when filters are returned to service. At present, there is no way to estimate in advance the effect of the above techniques for minimizing the initial turbidity spike, so a plant-specific trial-and-error approach is generally used to assess which technique might produce the desired filtered water quality. Some of the techniques described above require special equipment, such as chemical feed facilities for adding chemical to backwash water or filter-to-waste piping and valves for that method.

Modified Backwash Techniques

Adding coagulant to backwash water. Adding coagulant chemical or polymer to backwash water during the latter stage of filter washing is a technique that has been used for over three decades. When this is done, shut off chemical addition before ending the filter wash so the wash water in the underdrain and piping does not contain added coagulant or polymer, as this water will be the first water produced when filtration resumes. Generally, the chemical used would be a positively charged coagulant. The appropriate chemical dosage may need to be determined in pilot plant or full-scale plant filtration trials.

Extended terminal subfluidization backwash. Research by Amburgey et al. (2003) has shown the benefits of managing the backwash rise rate so that at completion of the usual backwash cycle, a subfluidizing rise rate is used to gently rinse backwash remnants (floc particles) out of the filter bed and the water above the filter bed before filter washing is concluded. The authors recommended that the subfluidized rinse should be done for a time sufficient to displace water in the filter bed and the water between the top of the media and the water surface. The objective of the gentle rinse after a vigorous backwash is to wash particles out of the filter bed without dislodging additional particles from filter media. This has been proven to be effective in full-scale plants. The authors reported that using a rise rate for the minimum subfluidization velocity for the effective size (d_{10}) of the filter media or for the d_{60} size of the media produced substantially better results than using a minimum fluidization velocity for the d_{90} size of the media in the filter bed. They noted that when filter media could be seen at the end of the subfluidization wash (i.e., the turbidity of the water over the media was low), the subfluidization wash had met the objective of removing backwash remnants and providing for lower turbidity when the filter was returned to service. This may be a useful preliminary indicator that the procedure was effective for plant operators who use the extended terminal subfluidization backwash.

Adding coagulant chemical to filter influent. Adding coagulant chemical or polymer to filter influent water as the filter box refills after backwashing is completed has been demonstrated to work with metal coagulants and with cationic polymer. Dosage is determined on a trial-and-error basis, but at two utilities the alum dosage (based on the volume of water in the filter box between the top of the media and the water surface) was about one-third of the alum dosage used for coagulation in pretreatment. If this technique is used with cold water, consider the reaction time of the coagulant in cold water and avoid discharging unreacted coagulant (such as alum) to the clearwell. This technique has been applied successfully on Lake Michigan water using cationic polymer.

Filter resting. Leaving a filter off-line after backwashing, for a period of time ranging from one half hour to 24 hr, has resulted in improved filter effluent turbidity at some plants. If this is done at utilities where excess filtration capacity exists, operators need to be careful to avoid leaving a filter out of service too long. Even if a free chlorine residual is carried onto a filter bed, the media will not be sterile but will have some attached bacteria. Metabolism by bacteria on filter media can deplete dissolved oxygen in the water in the filter bed, possibly causing taste and odor problems or redissolution of metals such as iron and manganese if anaerobic conditions occur. In a worst-case scenario, coliform bacteria might grow in the idle filter and cause coliform-positive samples in filter effluent.

Filter-to-waste.   Filter-to-waste is a widely practiced operational strategy to reduce the impacts of particle breakthrough after backwash. Simply, it involves wasting poorer-quality water from the filters for a period of time after the filter is returned to service. This wasting is a means of operating the filter to allow for improving filtrate quality without sending filtered water to the clearwell if it fails to meet the utility's quality goal. This method can involve a preset time based on experience that wasting should occur until particles reduce to acceptable levels. Operating in filter-to-waste mode for an insufficient time can result in discharging water to the clearwell even though the filtered turbidity does not meet treatment plant goals, whereas operating filter-to-waste too long results in wasting water that should have been sent to the clearwell. Therefore the preferable approach is to employ online turbidity analysis of the filter-to-waste water to enable operators to waste filtrate until the quality goal is attained. In some cases, up to several hours are necessary for wasting to achieve desired water quality, and opening and closing of valves can cause additional turbidity spikes due to rapid changes in flow (Amburgey et al. 2003). Controls should be provided to slowly and smoothly transition flow from filter-to-waste to filter effluent.

Slow start.   Starting a filter at a low filtration rate and gradually increasing the rate over time until the desired rate for operation during the run is attained is a procedure used at some plants. This approach requires filter rate control valves that can operate smoothly and with precise control, so they can be gradually opened a small amount, then left in position for a few minutes, then moved slightly again; in this manner, the filtration rate is gradually stepped up over time. This technique may be combined with filter-to-waste at plants with filter-to-waste piping that is inadequate for carrying the filter flow when the filter is operated in the upper range of its rate of flow.

Managing Filters During the Filter Run

During a filter run, filter management includes monitoring filtered water quality, monitoring head loss, and monitoring and controlling the rate of filtration. Monitoring the quality of filtered water was discussed in chapter 3, so the text below is focused on head loss and filtration rates.

Monitoring head loss is important because doing this enables operators to understand the status of each filter that is operating. If multiple filters reach terminal head loss in a short time period, this puts a strain on plant staff to perform backwashes, and removal of one or more filters for washing can result in the need to operate the remaining filters at higher rates, thus increasing head loss even more on the filters that have high head loss but remain in service.

Even though many rapid-rate granular filters are operated as so-called constant-rate filters, in fact filtration rates often are not constant throughout an entire filter run from start to finish. The need to increase or decrease water production and the need to remove filters from service for backwashing can result in application of rate changes to filters in service. Whereas decreasing the rate of filtration is unlikely to cause filtered water quality to deteriorate, increasing the filtration rate can do so. Cleasby et al. (1963) presented research results showing that filtered water quality could deteriorate when the filtration rate increased, with the extent of quality deterioration being greater when the rate increase occurred more rapidly, when the overall magnitude of the rate increase was greater, and when floc held in the filter bed was weak rather than strong. In addition, applying filtration rate increases can have a greater effect on filtered water quality when the filter is operating at high head loss, a condition indicative of filter pore spaces that are clogged with floc.

To counteract the potential for detrimental effects of filtration rate increases on filtered water quality, treatment plant staff should plan and manage filter operations to minimize the need for large or rapid increases in filtration rate. Filters that are close to reaching terminal head loss may need to be taken out of service for backwashing before a general rate increase is applied to all operating filters. Because operating filters without rate changes is not a realistic expectation, filter aid polymers, which often are added just prior to filtration to strengthen floc, may be needed. In a report of a survey of granular media filtration plants operating at a rate of 4 gpm/ft^2 (10 m/hr) or higher, Cleasby et al. reported that use of polymeric flocculation aids and/or filter aids was generally required (Cleasby et al. 1989). When high-molecular-weight polymers are used to strengthen floc, operators must be careful to apply an appropriate dosage. This is to some extent a balancing act, as if no filter aid or an insufficient dosage is used, weak floc can cause filters to be susceptible to turbidity breakthrough during filter runs. Using excessive dosages of polymer can result in higher rates of head loss accumulation in filter beds, with possible long-term problems related to backwashing and filter media condition. Either situation can cause filter runs to end prematurely.

Backwashing Filters

Filter backwashing procedures can influence filtered water quality. As explained above, some backwash procedures can improve the initial quality of filtered water when a filter is returned to service. In addition, long-term effects of inadequate backwashing can cause filtered water quality problems if mudballs form and the effective cross section of a filter bed is decreased, resulting in a higher filtration rate in the portions of the bed that are not clogged. Because increasing filtration rates on filters that have high head loss can promote turbidity breakthrough, spacing backwashes over the work day and paying careful attention to the head loss condition of each filter in service become important.

Another important aspect of filter backwashing that can substantially influence filtered water quality in an adverse way is the failure to backwash a filter that was in service for a period of hours, removed from service, and then returned to service with floc accumulated in the filter bed. When floc is weak or when a filter has high but not terminal head loss, restarting the dirty filter without backwashing puts great stress on the floc stored within the bed and can cause turbidity breakthrough. Restarting dirty filters may not be uncommon in small water systems where plant capacity is such that filters do not need to run continuously from return to service to attainment of terminal head loss or turbidity breakthrough, but this is a risky practice that should be avoided if possible.

At filtration plants that recycle spent wash water within the plant, returning the spent wash water to the head of the plant in large flows as backwashing occurs can result in changes in flow through the plant, which would necessitate changes in coagulant feed rates. And if recycling spent wash water changes the quality of raw water, a different chemical regime might be needed as compared to that for treating raw water with no wash water added. To avoid problems caused by large, intermittent flows of spent wash water, holding it in an equalization basin and returning it to the head of the plant at a more nearly uniform rate of flow can be helpful.

Importance of Properly Managing Filter Operations

Careful management and operation of all aspects of water filtration and filter washing are part of a program to optimize filtered water quality. Quality optimization begins with optimizing coagulation and pretreatment processes, but it must continue through filtration for the maximum benefits of treatment to be attained. Additional information on topics presented in this portion of chapter 5 can be found in *Water Filtration Practices* (Logsdon 2008).

REFERENCES

Ahmad, R., A. Amirtharajah, A. Al-Shawwa, and P. Huck. 1998. Effects of Backwashing on Biological Filters. *Jour. AWWA*, 90(12):62.

Amburgey, J., A. Amirtharajah, B. Brouckaert, and N. Spivey. 2003. An Enhanced Backwashing Technique for Improved Filter Ripening. *Jour. AWWA*, 95(12):81–94.

Amburgey, J., A. Amirtharajah, M.T. York, B. Brouckaert, N. Spivey, and M.J. Arrowood. 2005. Comparison of Conventional and Biological Filter Performance for *Cryptosporidium* and Microsphere Removal. *Jour. AWWA*, 97(12):77–91.

Amirtharajah, A., and D. Wetstein. 1980. Initial Degradation of Effluent Quality During Filtration. *Jour. AWWA*, 72(9):518.

Bouwer, E.J., and P.B. Crowe. 1988. Biological Processes in Drinking Water Treatment. *Jour. AWWA*, 80(9):82.

Cleasby, J.L., M.M. Williamson, and E.R. Baumann. 1963. Effect of Filtration Rate Changes on Water Quality. *Jour. AWWA*, 55(7):869–880.

Cleasby, J.L., A.H. Dharmarajah, G.L. Sindt, and E.R. Baumann. 1989. Design and Operation Guidelines for Optimization of the High-Rate Filtration Process: Plant Survey Results. Denver, Colo.: AWWA Research Foundation and AWWA.

Emelko, M.B., P.M. Huck, B.M. Coffey, and E.F. Smith. 2006. Effects of Media, Backwash, and Temperature on Full-Scale Biological Filtration. *Jour. AWWA*, 98(12):61–73.

Hozalski, R.M., S. Goel, and E.J. Bouwer. 1995. TOC Removal in Biological Filters. *Jour. AWWA*, 87(12):40.

Logsdon, G.S. 1979. *Water Filtration for Asbestos Removal*. EPA-60/2-79-206. Washington, D.C.: US Environmental Protection Agency.

Logsdon, G.S. 2008. *Water Filtration Practices*. Denver, Colo.: American Water Works Association.

Logsdon, G., J. Symons, R. Hoye, and M. Arozarena. 1981. Alternative Filtration Methods for Removal of *Giardia* Cysts and Cyst Models. *Jour. AWWA*, 73(2):111–118.

Mallevialle, J., I.H. Suffet, and U.S. Chan, eds. 1992. *Influence and Removal of Organics in Drinking Water*. Boca Raton, Fla.: CRC Press.

Manem, J.A., and B.E. Rittmann. 1992. Removing Trace-level Organic Pollutants in a Biological Filter. *Jour. AWWA*, 84(4):152.

Patania, N., J. Jacangelo, L. Cummings, A. Wilczak, K. Riley, and J. Oppenheimer. 1995. *Optimization of Filtration for Cyst Removal*. Denver, Colo.: AWWA and Awwa Research Foundation.

Rajagopolan, R., and C. Tien. 1976. Trajectory Analysis of Deep-Bed Filtration With the Sphere-in-Cell Porous Media Model. *Jour. AIChE*, 22(3):523–533.

Rittmann, B.E., and V.L. Snoeyink. 1984. Achieving Biologically Stable Drinking Water. *Jour. AWWA*, 76(10):106.

Robeck, G., N. Clarke, and K. Dostal. 1962. Effectiveness of Water Treatment Processes in Virus Removal. *Jour. AWWA*, 54(10):1275–1290.

Tobiason, John E., J.L. Cleasby, G.S. Logsdon, and C.R. O'Melia. 2011. Granular media filtration. In *Water Quality & Treatment*, 6th ed., J.K. Edzwald, ed. New York: McGraw-Hill and Denver, Colo.: AWWA.

Tufenkji, N., and M. Elimelech. 2004. Correlation Equation for Predicting Single-Collector Efficiency in Physicochemical Filtration in Saturated Porous Media. *Environ. Sci. Technol.*, 38 (2):529–536.

Urfer, D., P.M. Huck, S.D.J. Booth, and B.M. Coffey. 1997. Biological Filtration for BOM and Particle Removal: A Critical Review. *Jour. AWWA*, 89(12):83.

Young, G. 1961. *Whitten's Microbiology*. 3rd ed. New York: McGraw-Hill.

Chapter **6**

Pilot Testing for Process Evaluation and Control

Orren Schneider, James Farmerie, and Gary Logsdon

INTRODUCTION

Pilot testing is generally considered when a utility is contemplating making major capital improvements to a treatment plant. While desktop studies can screen possible technologies and bench testing can help identify possible treatment chemicals and dosage requirements, pilot tests give much more detailed design information regarding mixing energy, detention times, loading rates, and filtration. Without pilot testing, designs have to be much more conservative, leading to higher capital and operational costs to a utility.

Pilot testing requires significant resources, both in money and personnel. However, a properly conducted test will more than pay for itself in the long run because of lower design, construction, and operation and maintenance (O&M) costs. In a figure presented by Tate and Trussell (1982) to show the relationship between pilot testing costs and construction costs of plants for which testing was done, pilot plant costs tended to be in the range of 1 to 2 percent of plant cost when treatment plant costs ranged from about $1,000,000 to $50,000,000. In the years since 1982, an increased emphasis has been placed on performing expensive chemical and microbiolgical analyses for contaminants that were not a concern, or of which the water industry was not aware, in 1982. Because of higher analytical costs that can be associated with pilot plant testing in this era, total pilot plant study costs might be in the 2 to 3 percent range if a large component of laboratory analysis for disinfection by-products (DBPs), endocrine disruptors, protozoa, or viruses is a part of the study.

When possible, utilities should have their own personnel either conduct or participate in the testing to get operator "buy-in," to increase training and understanding of the new processes, and to allow the operators to contribute the utility's perspective in the design process.

Pilot testing is often performed by consulting engineers or by personnel provided by manufacturers of proprietary equipment or technologies. The use of utility personnel during testing (either for actual operations or analytical services) also helps to reduce external expenditures to defray some of the costs of piloting.

DETERMINING PILOTING GOALS

Establishing Goals

Pilot treatment goals will generally fall into one of two areas, water quality or economic. Economic goals may be to reduce capital or operational cost. This could include optimizing mixing time and energy and loading rate to reduce footprint or optimization of chemicals to reduce residuals volumes or increasing filter runs to minimize cost. Water quality goals may encompass any parameter of interest that the utility may have. With either water quality or economic goals, it is important to set forth the criteria for treatment early on in the planning process such that physical and experimental designs appropriate to the criteria can be made.

Typical performance goals will include the requisite water quality standards (although many utilities will want lower values during testing to allow for flexibility in operations) as well as production goals, often stated as unit filter run volume (UFRV) from each filter tested. Table 6-1 shows sample pilot treatment goals.

Type of Testing

Pilot tests generally fall in to two different categories: optimization or retrofitting of existing treatment plants, or testing of processes to replace existing facilities or build new greenfield plants. These two categories will necessitate different goals and requirements, leading to different piloting needs.

Optimization of existing processes. When attempting to pilot existing treatment plant processes in order to optimize treatment, the pilot plant must have hydraulic similitude to the existing plant. The first step in these types of pilot tests is to match the performance of the pilot plant to that of the existing plant so that any later optimization can be attributed to the changes in independent parameters (chemicals, mixing energy, loading rates, etc.) rather than to difference in hydraulics. In the best circumstances, identical parallel pilot trains will exist, both matching the existing treatment process. The second train can then be operated differently to examine the effect of changes.

Piloting new technologies. When examining major changes to existing treatment trains or examining new treatment facilities, parallel trains are not necessary. Once the decision has been made to use new processes, it is not necessary to have a comparison to the old (or nonexistent) process.

For replacement of old treatment facilities, results from the pilot plant can be compared to the existing plant provided that results are normalized, i.e., not by filter run length but by unit filter run volumes and so on. Water quality can be directly compared to determine the efficacy of the new process.

When treatment of a new water source is being evaluated, performance evaluations will be based on treatment goals, among which might be producing finished water having a quality similar to finished water from other treatment plants.

Table 6-1   Example pilot treatment goals

Parameter	Value	Associated Regulation or Goal
Turbidity	0.1 ntu	IESWTR, best practices
Particle counts	Steady state	Best practices
Ripening period	<20 min	Best practices
TOC	% removal based on raw TOC and alkalinity	Stage 1 D/DBPR
Algae	Maximize removal	Best practices
Aluminum	≤ raw water or 0.05 mg/L	Best practices
SUVA	<2.0 L/mg-m	Stage 1 D/DBPR
Iron	0.1 mg/L	State Sanitary Code
Manganese	0.05 mg/L	SMCL
Color	5 scu	State Sanitary Code
SDS TTHM	64 µg/L (80% of MCL)	Stage 1 and Stage 2 D/DBPR
SDS HAA5	48 µg/L (80% of MCL)	Stage 1 and Stage 2 D/DBPR
Tastes and odors	Maximize removal	Best practices
UFRV	>10,000 gal/ft^2 (about 400 m^3/m^2)	Economic considerations

Courtesy of Orren Schneider

NOTE: ntu—nephelometric turbidity units; IESWTR—Interim Enhanced Surface Water Treatment Rule; TOC—total organic carbon; D/DBPR—Disinfectants and Disinfection By-Products Rule; SUVA—specific ultraviolet absorbance; SMCL—secondary maximum contaminant level; scu—standard color units; SDS TTHM—simulated distribution system total trihalomethanes; MCL—maximum contaminant level; SDS HAA5—simulated distribution system sum of five haloacetic acids; UFRV—unit filter run volume.

Scale

The choice of flows will have significant impacts on the cost of pilot testing. Typically, the minimum flow is limited by the clarification technology, not filters or flocculation. If direct filtration is considered, flows may be as low as 5–10 gpm (19–38 L/m). However, for proprietary clarification technologies such as Superpulsator® clarifiers, Clari-DAF® dissolved air flotation system, or Actiflo® ballasted flocculation system, required flows may be in excess of several hundred gallons per minute. As the flow rate through the pilot increases, the data produced become closer to what one may expect from full-scale operations, therefore, uncertainty in design and full-scale operation is reduced.

The total flow to the pilot facilities should always be greater than the sum of the required flows to the different trains. While this does produce water waste, it also ensures that loading rates are never limited by supply. The wasted water (if not mixed with any residuals or treatment chemicals) can often be directed back to the raw water source.

Length of Testing

The longer testing can go on, the more useful the information produced. However, time on-site (for leased pilot plants, utility personnel, or consultants) has significant economic impacts on the utility. Some state regulators will require a minimum amount of time for testing, usually based on 6 or 9 months to cover multiple water quality seasons. Other states may require a year of testing for new technologies. For commonly used technologies, some states require only 2 weeks of testing.

Typically, if regulatory requirements are not an issue, testing should continue until previously determined water quality or design goals are achieved. Testing should be conducted during the period of time when the water quality is historically worst for the parameters of interest (often during spring or fall turnover for reservoir sources, or late winter to early spring for river sources) or when design flows will be highest.

When the pilot testing facility is owned by a water utility, a cost-saving option for pilot plant testing is to perform tests for several weeks on a seasonal basis or during times of challenging water quality. After sufficient data have been collected to establish performance capability for a certain water quality condition, further testing may not be needed until a change in water quality occurs. This can reduce the staffing burden for pilot testing, but staff must be flexible in their ability to conduct the testing when it is needed.

PROCESSES AND TECHNOLOGIES

Coagulation

Coagulation is the key to the treatment process. Without proper coagulation, the rest of the treatment train will not work properly. During pilot testing for new processes, it is important to test multiple chemical combinations, order of addition, and dosages to assess effects on treatment as well as to determine usage rates to set design criteria for pumps and chemical storage.

Chemical addition.   Flows in pilot plants are typically much lower than full-scale facilities, so flows of chemicals are consequentially much lower. These low chemical flows introduce challenges to pilot plant design and operations. For instance, when using full strength alum (48 percent) with a dosage of 20 mg/L in a plant flow of 20 gpm (76 L/m), the required flow rate would be 2.4 mL/min. Most standard chemical feed pumps cannot pump at this rate or require such low rates as to cause pulsing and poor dispersion of the chemicals.

Alternatives to standard chemical metering pumps are available. Peristaltic pumps are available with wide ranges of flow rates. When using peristaltic pumps, it is necessary to make sure of chemical compatibility between the chemical being pumped and the tubing material. It is also advisable to inspect the tubing to ensure that undue wear is not occurring.

A second alternative for chemical addition is to dilute the full-strength chemicals to achieve a more reasonable flow rate. For coagulants such as alum, ferric salts, or polyaluminum chloride, this is not preferable as hydrolysis may begin to occur. For acids, bases, organic polymer, and other chemicals, dilution is acceptable as long as fresh chemical solutions are replaced daily.

Addition of chemicals into pipes can also create difficulties during piloting. Because chemical feed lines are often not hard-piped, air bubbles tend to accumulate in bends and can result in air binding and improper chemical feed. In order to minimize this, the length of feed lines should be minimized, and chemicals can be fed into the bottom of pipes to allow any air bubbles to escape from the tubing into the piping.

If slurries are fed, sloping the feed line downhill from the pump to the point of addition helps avoid clogged feed line problems.

Mixing intensity and time.   The mixing intensity during coagulation should be the same as that of the full-scale system. For optimization tests, this is relatively easy to determine. For new processes or plants, flexibility should be planned into the pilot program to allow for testing of several different combinations of intensity and time.

Flocculation

Hydraulic similitude.   It is of utmost importance that the design of pilot flocculation facilities be based upon the full-scale design. For simulations of existing facilities, length-to-width ratios and contact times should mirror the existing plant. For facilities yet to be built, the full-scale design should be based on the pilot facilities. Baffling of pilot flocculation basins (Figure 6-1) is also important to prevent short-circuiting.

Mixing time.   The flocculation mixing time used should be appropriately matched to the clarification process. For direct filtration, dissolved air flotation, and sand-ballasted systems, the mixing time may be as low as 5 min at the design flow. For conventional sedimentation processes, the flocculation time can be as high as 30 min or more.

Mixing energy.   As with flocculation mixing time, it is important to match mixing energy with the clarification process being tested. In order to create "pin" flocs, higher flocculation energies are generally used. For sedimentation processes, larger flocs are desired, leading to lower mixing intensities to prevent excessive shear forces.

During pilot testing, it is important to measure the power draw of the flocculator motors and to measure the rotational speed of the mixing blades in order to assess the velocity gradient, G, imparted to the water. This factor will be an important design parameter.

Clarification Processes

Clarification processes include conventional sedimentation basins, basins equipped with tube settlers or plate settlers, solids contact clarifiers, ballasted flocculation, and dissolved air flotation. In these processes, removal of organic carbon will be limited by

Source: Schneider et al. 1998.

Figure 6-1   Example of pilot-scale flocculation basins

the coagulant type and dosage, pH, alkalinity, and type of organic matter present. The organic carbon removal criterion should be based on the requirements of the Stage 1 Disinfectants and Disinfection By-Products Rule (D/DBPR).

Clarification is not the ultimate drinking water treatment but instead serves to remove particles and floc from water so water quality can be further improved and filter runs can be longer. Pretreatment can be managed to maximize clarification performance, but this will not necessarily result in the best filter performance. For example, using greater polymer dosages may result in very low clarified turbidity, but the high polymer dosage then could cause shortened filter runs. Clarifiers and filters work together in a full-scale plant, so pilot testing of clarification processes should be followed by filtration of the clarified water to understand the combined results of clarification and filtration.

Conventional gravity sedimentation. Because of its low loading rates, piloting of conventional gravity sedimentation is rarely performed. In instances where comparisons to an existing sedimentation basin are desired, the pilot basin should be designed with a similar length-to-width ratio and the same depth as the full-scale system.

A typical performance criterion for the sedimentation portion of the pilot is a clarified turbidity less than 2 ntu. Because of the extra footprint required for sedimentation, a typical economic criterion includes a UFRV of greater than 10,000 gal/ft^2 (about 400 m^3/m^2).

In addition to water quality information, important measurements to collect during piloting include information about residuals: percent solids, dewatering requirements, volume, and so on.

Tubes/plates. When utilities want to upgrade older conventional treatment plants, tube or plate settlers are attractive alternatives that can maximize existing infrastructure. Because sedimentation is involved, higher coagulant dosages and flocculation times are typical for tube or plate sedimentation units than for some other high-rate clarification processes such as DAF and contact adsorption clarifiers.

As with full-scale installations, when tubes or plates are being piloted, care must be taken to avoid bridging of solids between the walls. If bridging does occur, effective flow area is diminished leading to artificially high loading rates in some parts of the basins. Additionally, if the bridging solids are sloughed, a relatively high mass of solids will be deposited on the filters leading to poor performance.

Solids contact clarifiers. Solids contact clarifiers (Figure 6-2) use combined flocculation and clarification processes. As coagulated particles pass through the sludge blanket, they flocculate and attach in the sludge layer, effectively being clarified.

Because of scale-up issues, pilot-scale units are generally virtually full-sized versions of the technology without associated filters. As such, flow rates to these "pilot" units can be several hundred gallons per minute.

During pilot testing, performance can be affected by diurnal variations in temperature. As the temperature varies, the density of the blanket can change. If the blanket becomes buoyant, the blanket can lose its integrity and pass from the clarifier onto the filters, resulting in poor water quality and performance. Another issue to monitor during testing is to ensure that the sludge blanket does not go anoxic. If biological activity is present in the sludge and anoxia occurs, oxidized metals such as iron or manganese can be released and pass onto the filters. Without an oxidant, these now reduced metals will pass through the filters. When measuring these metals as part of the testing, high results could lead to erroneous conclusions about oxidant demands, leading to poor designs.

A typical pilot testing performance criterion for a solids contact clarification process is a clarified turbidity less than 1 ntu. Because of the extra footprint required for

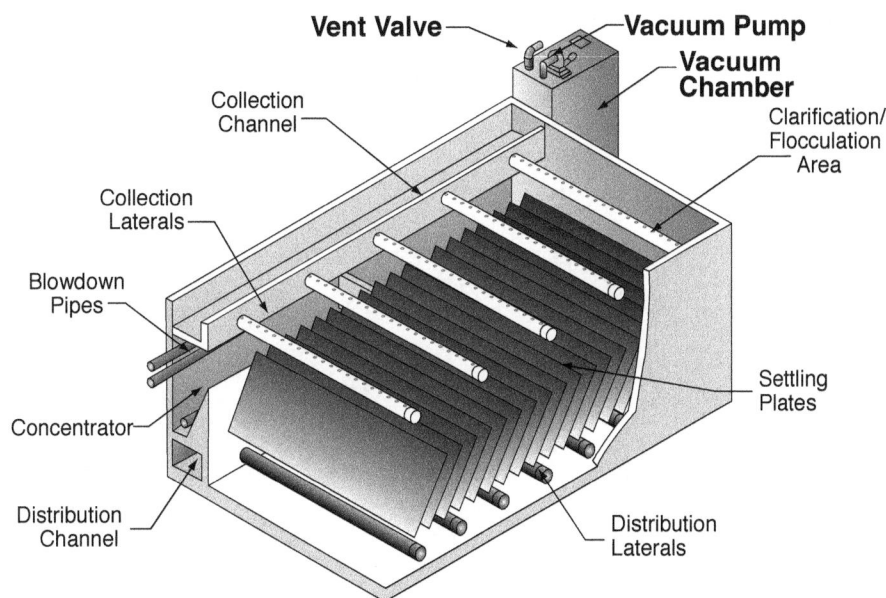

Source: Degremont Technologies.

Figure 6-2   Schematic of the Superpulsator® solids contact clarifier process

clarification equipment, a typical economic criterion includes a UFRV of greater than 10,000 gal/ ft^2 (about 400 m^3/m^2).

In addition to clarified water quality and chemical usage, important design parameters to determine during piloting are sludge blowdown frequency, blanket thickness, and information regarding residuals (percent solids, dewatering characteristics, volume, etc.).

Ballasted flocculation.   Ballasted flocculation pilot plants (Figure 6-3) are basically small full-scale systems without filters. As with the full-scale process, selection of the polymer and inorganic coagulant is the key to proper performance. Because the ballast, when attached to the floc, is principally responsible for sedimentation, the denser floc size is not as critical to performance as it is with other processes. However, smaller flocs work well with ballasted flocculation and therefore allow for smaller floc basins and lower chemical dosages than those required for sedimentation.

Critical information to determine during piloting is the coagulant dosage, selection of the polymer and its dosage, and clarification loading rate.

A typical performance criterion for the ballasted flocculation portion of the pilot is a clarified turbidity less than 1 ntu. Because of the extra footprint required for the clarification section of the ballasted flocculation, a typical economic criterion includes a UFRV of greater than 10,000 gal/ ft^2 (about 400 m^3/m^2).

For states where ballasted flocculation is not widely used, it may be important to work with regulators to establish treatment goals and other piloting requirements. It may be useful to conduct microbial removal studies to demonstrate the capabilities of this technology.

DAF.   Piloting of dissolved air flotation (Figure 6-4) can give excellent information regarding this process if the pilot systems are designed correctly. Because pressurized water flow is one of the relatively high costs of power for this process, it is important that the air/water recycle stream is properly dispersed into the main flow. This

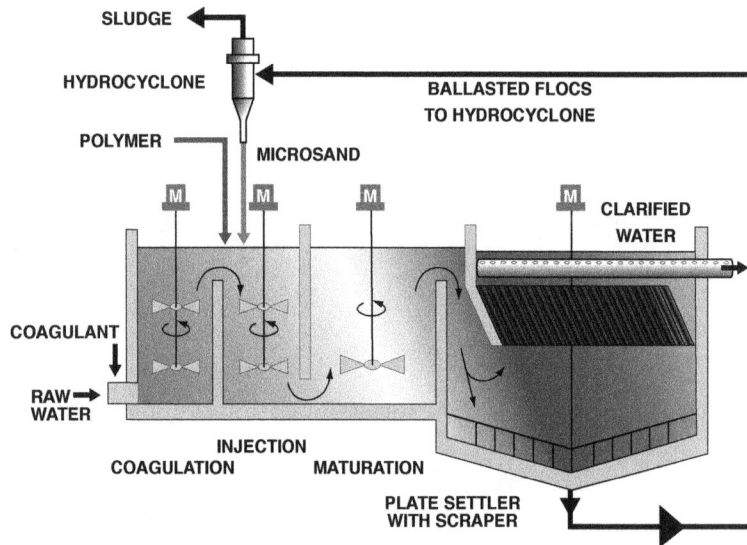

Source: I. Krüger Inc., A Veolia Water Solutions & Technologies Company.

Figure 6-3   Schematic of the Actiflo® ballasted flocculation clarification process

dispersion is easier to accomplish at higher flows, thus, based on conservative loading rates (4 gpm/ft^2 or 10 m/hr), the minimum flow to a DAF pilot should be on the order of >30 gpm (>110 L/m), and higher loading rates (16 gpm/ft^2 or 40 m/hr) would require flow to the DAF pilot on the order of >120 gpm (>450 L/m).

Because DAF works by lifting the floc to the surface rather than allowing it to settle, the type of floc required is very different from that required for sedimentation. For this process, "pin" flocs, similar to those needed for direct filtration, are desirable. These types of flocs can be achieved with relatively low coagulant dosages and short flocculation periods. If high dosages of coagulant are required for sedimentation processes in order to settle algae, then coagulant dosages for DAF can indeed be lower. Or where Fe/Mn are oxidized out of solution, typically less chemical is required to coagulate the small precipitated particles to form a floc large enough to remove. However, if dissolved organic matter and typical turbidity-causing particles are the main source of coagulant demand, the coagulant dosages used in DAF will be very similar to those required for sedimentation processes.

Based on anecdotal observations, when the bubbles in the sludge become smaller, coagulation is getting better. Conversely, large bubbles may indicate problems with the saturation system with either the air compressor psi or the valve in the reaction tank not properly functioning.

A typical performance criterion for the DAF portion of the pilot is a clarified turbidity less than 1 ntu. Because of the extra footprint required for DAF, a typical economic criterion includes a UFRV of greater than 10,000 gal/ft^2 (about 400 m^3/m^2).

In addition to clarified water quality and chemical usage, important design parameters to determine during piloting are chemical(s) and dosage(s), order of addition, flocculator mixing energy and time, loading rate, air usage and recycle rate, saturator pressure, and information regarding residuals (percent solids, percent of forward flow, desludging frequency, dewatering characteristics, volume, etc.).

For states where DAF is not widely used, it may be important to work with regulators to establish treatment goals and other piloting requirements. It may be useful to conduct microbial removal studies to demonstrate the capabilities of this technology,

Source: ITT Water & Wastewater.

Figure 6-4   Example of a pilot-scale DAF basin with scraper-type sludge removal

although because of the cost involved in such studies, documenting DAF experience in other jurisdictions may be a more cost-effective means of gaining regulatory approval.

Filtration

Following clarification, water is filtered. Factors considered in pilot plant tests of filtration include media design (mono-medium versus dual-media or multimedia, with variables including the depth and effective size of each layer of filter material), filtration rate, and, of course, filter performance. The latter is assessed in terms of filtered water quality, total water production in a filter run, or UFRV expressed as gal/ft^2 (or m^3/m^2), and rate of accumulation of head loss. Some pilot filters are designed with piezometer taps in the filter column so head loss development at different depths within the filter bed can be measured. This information typically is collected only occasionally and not on a continuous basis as with total head loss. The piezometer taps are useful for learning where floc is being deposited within the bed during the run and causing the head loss to increase. Very strong floc resulting from an excessive dosage of filter aid tends to be removed in the upper reaches of the filter bed. With modern filter bed designs, floc removal within the bed is desired as a way to distribute head loss more uniformly down into the bed and attain longer filter runs. Another approach to evaluating filter performance is to continuously withdraw a small sample stream from within the filter bed, for example at the interface of anthracite and sand in a dual-media filter, and then measure the turbidity or particle count in that sample and compare those results with results for water that has passed through the entire filter bed. When preceded by clarification, filters should produce 10,000 gal/ft^2 (about 400 m^3/m^2) or more.

Direct filtration is a special case involving filtration in which no clarification process is used prior to filtration. The use of direct filtration has been declining in recent years due the recognition that clarification processes act as an additional barrier to pathogens. Nonetheless, for appropriate raw water qualities, direct filtration can be a feasible alternative. Because no clarification step is included, the floc required for direct filtration is much different than that required for sedimentation. Additionally, because the filters act as the sole particle removal mechanism, proper chemical addition (coagulant type and pH), mixing energy, and mixing time are critical to proper performance. Typically, the range of proper chemical dosages is smaller than for other processes. Direct filtration pilots should examine several media configurations and loading rates in addition to chemical dosages and mixing times and energies.

Apart from filtered water quality, typical economic criteria include a UFRV of greater than 5,000 gal/ft^2 (about 200 m^3/m^2). This criterion can be lower than for process trains with a clarification step, as the additional cost of residuals treatment is offset by the reduced plant footprint.

Critical piloting information to collect includes chemical dosing, mixing energy, filtered water production, and analyses of waste filter backwash water (total suspended solids [TSS] and other measures) so that properly sized residuals treatment facilities can be designed.

INSTRUMENTATION

Instrumentation for pilot plants (Figure 6-5) can run the range from simple grab sampling to full automation. If pilot tests are to be run unstaffed overnight, provisions should be made to have data recorders and/or acquisition systems available to collect and store the data produced. This extra cost of data collection nonetheless maximizes the capital investment in the pilot plant by maximizing the time that the pilot plant is being used.

For the most part, instruments used during pilot tests are identical to those used in full-scale operations. If pilot facilities are not supplied with instruments, if a utility has spare or relatively recent obsolete equipment, these can be used once properly calibrated.

Turbidimeters

Turbidimeters should be used to monitor water quality in the raw water, clarified water, and treated water. For filtered water, turbidimeters can either be dedicated to individual filters (as shown above) or, with proper controls and purging, one turbidimeter can be used to measure turbidity from multiple filters.

Source: Schneider et al. 1998.

Figure 6-5   Example of pilot filter gallery showing dedicated turbidimeters and differential pressure transmitters sending data to a central data acquisition system

For applications with high solids concentrations (such as monitoring backwash water turbidity), surface scatter turbidimeters can be used to improve accuracy.

Differential Pressure

Differential pressure monitors are important for good pilot tests to collect head loss data. If older style piezometer boards are used, the pilot plant must be staffed whenever the pilot plant is operated to ensure accurate head loss data. If automated systems are used, data can either be transferred from recorder charts or logged automatically.

Differential pressure transmitters should be placed just above the top of the filter media and below the underdrain in order to reflect head loss across the filter.

The pressure probes should be periodically recalibrated to ensure accuracy. One way to check their calibration is through the use of piezometer tubes.

pH

In order to optimize treatment, measurement of pH is very important. The raw water and coagulated (or flocculated) water should be monitored so that final designs can be improved. Most coagulants have an optimum pH range for effectiveness. In addition, numerous utilities are looking at "enhanced coagulation" for organics removal in clarification, and pH is a key factor in use of metal coagulants for control of organics. Finished-water pH should be monitored to provide data related to the possible need for corrosion or scale control for the distribution system. Because pH probes tend to foul because of raw water solids or floc, the probes should be regularly cleaned and calibrated.

Flow Control

In order to determine process loading rates, flow control is important. While rotameters equipped with valves can set flow, variations in pressure or tank elevations can lead to variations and flow surges. In order to accurately measure and record flow, automatic flow monitors should be used. Knowing the rate of flow in any process being studied is very important. Knowledge of raw water flow into the treatment train is necessary for calculation of chemical dosages. This information is also essential for knowing overflow rates in clarifiers. Knowing the rate of flow out of each filter is a must for determining filtration rate. A quality control check on flowmeters of any type is to collect flowing water for a measured period of time, measure the volume, and, using time and volume data, determine the rate of flow in gpm or L/min.

Particle Counters

Particle counters can be useful for process optimization. While many people think of using particle counters on raw and finished water to determine log removal, this is probably not a good application. Instead, particle counters can be better used to understand filter ripening and turbidity breakthrough of filters. Particle counters can also be used to get a sense of steady state operation of filters. Because of high numbers of particles formed by coagulation, the use of particle counters in coagulated and even clarified waters is not suggested.

If particle counters are used during pilot tests, either bench-top or flow-through types can be used. Prior to starting a pilot test, the counters should be recalibrated by the manufacturer.

Streaming Current Monitor and Zeta Meter

In order to optimize coagulation, either streaming current monitors (SCMs) or zeta meters can be used. While not critical to operation or design, having an electrokinetic charge analyzer can aid in optimization if different coagulants and pH ranges are tested. When regularly used, these instruments can signal when significant changes in raw water quality occur or when coagulation is not optimized.

QUALITY CONTROL

Collection of valid data during the conduct of a pilot study is essential. The quality of data cannot be ensured unless a program of quality control is carried out through the duration of the study. This applies to data obtained by chemical and microbiological methods and also to physical and operational data collected during the study. Neither type of data (analytical data or the physical and operational data) is more important than the other kind. Strong emphasis often is placed on quality control for analytical procedures and data, and this is appropriate. The importance of obtaining valid physical and operational data must not be overlooked.

Examples of inadequate quality control for physical or operational data include a state drinking water regulatory engineer stating that a package water treatment plant evaluation had been done without any rate of flow data being collected. In the absence of flow rate data, filtration rates and basin retention times could not be determined. In a report on a year-long demonstration-scale evaluation of reconfigured filters, media, and an improved underdrain at a large plant in the western United States (reviewed by one of the authors), one filter was operated for the duration of the study with head loss instrumentation that indicated a clean bed head loss as high as 6 ft (1.8 m) when filtering at 8 gpm/ft^2 (20 m/hr) as contrasted to a 2.5-ft (0.76-m) clean bed head loss for another filter operating at that rate. At the end of the study, it was concluded that the unusually high head loss value was an instrument artifact. Checking the validity of head loss data at the first sign of a large difference in the instrument readings would have been the appropriate action to take.

With the trend of ever-increasing reliance on instruments for continuous monitoring of water quality and physical operating conditions, maintaining a vigilant quality control program during pilot testing is perhaps even more important than it was when much of the analytical work was done by analysts who tested grab samples. Frequent review of testing results is necessary in order to identify strange results or outliers. When problems seem to be happening with analytical or monitoring methods, acting immediately to identify that a problem really does exist can save much wasted effort when erroneous data are being collected. However, if quality control efforts confirm that data are valid, then continued operation and testing can be carried on with confidence. Learning of analytical or monitoring problems at the end of a pilot plant study can call the entire study into question, so it is much better to maintain a strong quality control program and minimize the potential for collecting spurious data.

SPECIAL TESTING

While pilot testing should be planned for worst-case conditions, including water quality, nature does not always cooperate. Prior to starting the pilot testing, contingency plans should be in place to artificially spike compounds of interest should the "right" raw water conditions not arise.

Taste and Odor (T&O) Spiking

If tastes and odors are a concern to a utility and pilot testing includes oxidation or adsorption processes, the use of surrogates such as geosmin or 2-methylisoborneol (MIB) should be considered. These compounds can be injected into the raw water and allowed to run through the process train. Samples can be collected and analyzed by using gas chromatography (GC) or gas chromatography/mass spectroscopy (GC/MS). Geosmin and MIB are expensive to purchase and analyze for, so proper planning and execution are important to keep costs down.

These compounds are typically not well removed by coagulation and clarification. If oxidation and/or adsorption are part of the pilot, then taste and odor testing can be appropriate. If these processes are not used, then it is not logical to spike for tastes and odors. As some of the chemicals that may be spiked in a pilot test are expensive, bench-scale prior to pilot testing can be a way to identify preferred approaches to use in larger-scale (pilot) tests, if work at that scale is needed.

Microbial Spiking

While tastes and odors may commonly be found in raw water supplies, it is less common to encounter large numbers of pathogens in the raw water. In order to test treatment processes under extreme conditions and determine the process train's ability to achieve performance goals, it is sometimes desirable to spike large numbers of organisms including *Giardia* cysts, *Cryptosporidum* oocysts, and MS-2 viruses. When using these organisms, care must be taken to ensure that the water and wastewater are properly disposed of to prevent introduction of live, infectious organisms to the environment (especially with the *Cryptosporidum* oocysts). As with the taste and odor studies, only a limited number of microbial spiking experiments should be run. Proper planning and execution are required to keep costs down. An alternative to spiking microbes in turbid surface waters is to test for *Bacillus* spores in raw and treated water, as these are found in surface runoff.

Turbidity Spiking

To examine the effects of a high-turbidity runoff event on a process train's treatment ability, it is possible to artificially increase the turbidity by spiking the water using soil, mud, or sediment gathered from the watershed. This material can be fed into the raw water line to achieve a desired turbidity. As this material will also have some organic matter associated with it, it becomes a good method for estimating impacts of runoff on treatment performance and requirements.

Other Tests

When water quality parameters are of interest, e.g., arsenic, but may not be present in the water because of seasonal variations or lower than normal background levels, these compounds can be spiked into the water either as a step change or as a single short-duration spike. The type of test will depend on the parameter of interest as well as the economic and physical feasibility of performing the test. These decisions should be considered during the planning stages of the pilot testing program.

EVALUATION OF TESTING RESULTS

Evaluation of testing results is an ongoing activity during pilot plant testing. Waiting to evaluate results of a test program until testing is completed is very risky, because discovery of serious deficiencies in pilot operations or sample analysis at the end of testing could invalidate weeks or months of work. Operating a pilot testing program for weeks and weeks while using incorrect coagulation chemistry and obtaining poor treatment results is an exercise in futility.

The performance of clarification processes and of filters is related to coagulation chemistry. The ultimate proof that the right coagulant, the correct dosage, and the proper pH are being employed is the quality of water produced in clarification and filtration. These results should be assessed every day by the pilot plant operating staff during testing and periodically by the managers of pilot investigations.

It is especially important to look for the unexpected or odd result that does not seem appropriate, and to follow up to learn if the work has been done correctly and unexpected results are valid, or if some problem developed but was not detected, causing results that are not valid. Sometimes an outside review by senior staff may be needed to point out the obvious and correct a trend of unsuccessful testing. Because source water quality conditions can change during a testing period, major reviews of results should be undertaken following periods of high algae concentrations, high turbidity, high concentrations of natural organic matter (NOM), or other water quality episodes that might cause treatment difficulties. Because of the transient nature of some water quality episodes, operating staff need to be diligent to identify effective coagulation chemistry as quickly as possible so time is available for confirmatory pilot plant runs during the episode.

Following completion of the testing and data analysis, it is important to use an objective, rational method to evaluate the different process options. This method can range from as simple as ranking the different processes on a qualitative basis to more elaborate methods involving statistical analysis.

It is often helpful to break down the evaluation criteria into a number of different categories including both numerical (water quality, economic) and qualitative (functional issues, flexibility, etc.) Table 6-2 shows sample evaluation criteria for a hypothetical pilot testing program with three process trains.

The rating of the process trains can be done individually by the various stakeholders (utility managers, plant operations, consultants, regulators) and then a workshop held to reach consensus on the final ratings for the different categories and criteria. Once the final ratings have been given to all of processes, an overall score can be given and compared against preliminary costs. Then a decision can be made regarding future plans.

Using Pilot Filters as Online Monitoring Tools

Another use for small-scale filters, in addition to their use in pilot plants, is for monitoring efficacy of coagulation as practiced in the full-scale plant. Filtration plant operators can use pilot filters together with an online turbidimeter as an online monitoring system to assess turbidity of water that has been coagulated and filtered and thus indicate whether coagulation is appropriate for good particle removal. Generally the filter columns are 4 to 6 in. (10 to 15 cm) in diameter. To be good indicators of full-scale filter performance for removal of turbidity-causing particles, pilot filters need to have the same media design as the filters in the plant. Turbidity of pilot filter effluent is continuously monitored to provide "after-the-fact" information on filterability of the coagulated particles formed by chemical addition and mixing. A typical pilot filter application consists of a pair of filter columns that are used alternately. During operation of a pilot filter, particles are removed in the filter bed, which gradually clogs. When terminal head loss is reached, that filter is removed from service and the other pilot filter is placed into service. Then the clogged pilot filter is backwashed. Use of pilot filters in an alternating pattern ensures that one filter is always operating and monitoring filtration efficacy. The filtration rate for the pilot filter should be similar to

Table 6-2   Example evaluation criterion for a hypothetical pilot testing program

Category	Criterion	Process Train A	Process Train B	Process Train C
Water quality	Disinfection by-products	++	+	−
	Pathogens	++	+	+
	Finished water stability	−	+	++
	Aesthetics (Fe, Mn, T&O, hardness)	++	++	+
	Meet future regulations	++	++	++
	Other (As, perchlorate, etc.)	NA	NA	NA
Functional issues	Long-term process reliability	+	+	+
	Maintenance complexity	− −	−	+
	Operations complexity	− −	−	+
	Safety	−	0	0
	Unattended operation	− −	−	++
	Staff skills	− −	+	+
Environmental	Noise	−	+	−
	Traffic	0	0	0
	Public safety	0	0	−
	Water recovery (backwash)	++	+	−
	Residuals	+	+	−
	Footprint	− −	−	+
	Visual impact	+	+	+
	Chemical usage	−	− −	+
	Water rights	NA	NA	NA
	Construction impacts	+	−	− −
	Constructability energy	+	+	+
Flexibility	Treat changing water quality	++	+	+
	Expansion capability	− −	+	++
	Treat diurnal fluctuations	++	+	++
	Treat seasonal fluctuations	++	+	−
Economic	Capital			
	O&M			
	Life cycle			

++ best; + good; − poor; − − worst; 0 neutral; NA not applicable

Courtesy of Orren Schneider.

that used at full-scale, but varying the rate on pilot filters to keep up with filtration rate changes in the plant may not be practical.

The filter influent for pilot filters is a sidestream of coagulated water that has been discharged from the rapid mix before any kind of clarification can take place. A continuous stream of water is extracted from the plant at a location that provides a water sample representative of that going to the flocculator, and sometimes this sample is pumped to the pilot filter installation. Floc can be damaged when a centrifugal pump is used. Therefore, if this type of pump is used to feed water to a pilot filter, the water sample must be extracted from the plant after coagulant chemical is thoroughly dispersed but before floc forms.

Even though pilot filters simulate a full-scale in-line filtration process train, employing only coagulation, rapid mixing, and filtration, they can be used to evaluate coagulant dosage for source waters with a much wider range of turbidity than would be considered treatable in a full-scale plant employing no clarification process. They have been used successfully on source water in Oregon even when the turbidity rose to 1,400 ntu and higher during a major flood event in 1996 as described in *Journal AWWA* (Wise 1998). When the turbidity was over 1,400 ntu and the plant was operated in a filter-to-waste mode, the pilot filter apparatus was used to evaluate the efficacy of the coagulant dosages being tried before a final dosage was set and filtered water was sent to the clearwell (D. Wise, pers. comm., 2008). Pilot filters are reported to produce filtered turbidity that typically is slightly lower than the combined filter effluent turbidity. The pilot filter clogs more quickly when turbidity is high and coagulant dosages are high, but because evaluation of filter run length is not an objective of operating this kind of filter, it is simply taken off-line and the other filter is placed online so the clogged filter can be backwashed.

The key filter performance indicator is effluent turbidity. Head loss and rate of flow also are monitored so the operating conditions of the filter are known. The turbidity of filtrate from a pilot filter with the same media design as that used in full-scale filters provides plant operators with an early indication of the filter effluent turbidity after flocculation, sedimentation (if the latter is used), and filtration in the full-scale process train. At conventional plants, the time of travel between entry to flocculation and appearing in filter effluent can be several hours, but the time between coagulation and passage through a pilot filter would be a fraction of an hour. If for some reason coagulation suddenly is not effective, having a pilot filter online is a way for the plant operator to obtain an early warning that a problem exists, perhaps hours before a conventional settling basin is full of water that is not going to be filtered effectively. Pilot filters also can quickly confirm that a change in coagulation practice indicated by jar test results or streaming current readings is actually going to be effective. Thus, pilot filters can verify the validity of chemical pretreatment decisions made using other approaches and provide additional evidence of filterability of the pretreated water.

Pilot filters cannot be used to predict the rate of head loss accumulation, and they cannot be used to provide guidance on filter aid dosage, as they treat water that only has been coagulated. These filters are not capable of predicting filter run length in the plant. However, this is not an obstacle to the use of pilot filters, because other procedures used for assessing pretreatment chemistry also fail to provide information on filter run length.

Maintenance activities for pilot filters include periodic checking of piping and valves, verification that media have not been lost from the filters during backwashing, and checking on the condition of the pump, if one is used. Quality control checks include periodic verification of the effluent turbidimeter's calibration and verification that measurements of head loss and flow rate are accurate. These filters are sold commercially, and they also can be fabricated by personnel at water treatment plants.

REFERENCES

Schneider, O.D., A.D. Nickols, J.K. Schaefer, and W. Kurtz. 1998. The Use of Ozone and Biofiltration to Meet Simultaneous Treatment Goals. In *Proc. Water Quality Technology Conference, San Diego, Calif.* #3B4. Denver, Colo.: AWWA.

Tate, C., and R. Trussell. 1982. Pilot Plant Justification. Presented at 24th Annual Public Water Supply Engineers Conference, Champaign, Ill., April 27–29, 1982.

Wise, Doug. 1998. Cross-Training Benefits Oregon Plant. *Jour. AWWA,* 90(10):60–66.

This page intentionally blank.

Chapter **7**

Case Studies

This chapter consists of nine case studies. Four provide information on applications of online instrumentation for process control in water treatment plants. Two case studies describe results of jar test programs. Two others focus on changes in coagulant chemical for improved water treatment and in polymer use for improved sludge handling and management. One presents a statistical approach for analysis and interpretation of turbidity data.

Titles and authors of the case studies in the order in which they are referenced in the first four chapters of this manual and are presented in chapter 7 are:

- "Conversion From Alum to Ferric Sulfate at the Addison-Evans Water Treatment Plant, Chesterfield County, Va.," George Budd and George Duval, cited in chapter 1

- "Jar Test Calibration," George Budd and Paul Hargette, cited in chapter 2

- "Relationships Between Coagulation Parameters, Winston-Salem, N.C.," George Budd, Paul Hargette, and Bill Brewer, cited in chapter 2

- "NOM Measurements for Coagulation Control," Tom Elford and David J. Pernitsky, cited in chapter 3

- "Net Charge Equals Positive Change," David Teasdale, cited in chapter 3

- "Streaming Current Detector Pilot Study: The Detection of a Ferric Chloride Feed Failure," Michael Sadar, cited in chapter 3

- "The Application of Simplified Process Statistical Variance Techniques to Improve the Analysis of Real-Time Filtration Performance," Michael Sadar, cited in chapter 3

- "Online Monitoring Aids Operations at Clackamas River Water," Robert D. Cummings, cited in chapter 3

- "Palm Beach County Water Utilities Water Treatment Plant 8 Ferric Chloride Addition," Tim McAleer and Jose Gonzalez, cited in chapter 4

These studies are intended to present illustrative examples that can be adapted to other water treatment facilities by users of this manual and to demonstrate how the concepts in some of the previous chapters can be applied.

$$\begin{matrix} Case \\ Study \end{matrix} \mathbf{1}$$

Conversion From Alum to Ferric Sulfate at the Addison-Evans Water Treatment Plant, Chesterfield County, Va.

George Budd and George Duval

In the late 1990s, coagulation evaluations were conducted at the Addison-Evans Water Treatment Plant (WTP) in Chesterfield County, Va., in an effort to improve organics removal and decrease the levels of disinfection by-products (DBPs) in the finished water. This plant contains a conventional rapid mix, flocculation, sedimentation, and filtration sequence with a design flow of 12 mgd (45 ML/d). Source water is obtained from a reservoir that is characterized by low turbidity and alkalinity and relatively high total organic carbon (TOC). Evaluations that were performed included multidimensional coagulant characterizations that involved bench-scale testing of alum, ferric chloride, ferric sulfate, and polyaluminum chloride to identify the regions of best performance for both particle and organic removal. As a result of the coagulation evaluations, a full-scale conversion from alum to ferric sulfate commenced in early 1999. Results of the full-scale conversion were reported by Budd et al. (2004).

Figure 7.1-1 illustrates the change in TOC removal that resulted from the full-scale conversion to ferric sulfate as the primary coagulant. Raw water quality conditions during the time period evaluated (January–September 1999) were characterized by alkalinity of 7–10 mg/L as $CaCO_3$, turbidity of 9–12 ntu, pH of 6.7–6.9, and raw water TOC concentrations of 4.5–7.0 mg/L. At the time of conversion to ferric sulfate, a detailed ferric sulfate coagulant profile for settled turbidity was developed from bench-scale results. The detailed profile is contained in Figure 7.1-2, and indicates an optimum pH for coagulation with ferric sulfate in the range of 5.4–6.0, with a ferric sulfate dose of 30–40 mg/L. Full-scale operation with ferric sulfate in the pH of range of 5.2–5.5 was subsequently found to provide effective floc formation for clarification and filtration during the initial period following the conversion. Although the change in coagulant chemical was driven by a goal of improved TOC removal, it was essential to utilize coagulation conditions which provided effective floc formation as well as increased TOC removal.

For the Addison-Evans WTP, the pH for effective operation with ferric sulfate has remained in the same general range since the conversion to ferric sulfate. Ferric sulfate dosage has been increased when required by changes in water quality that can occur following storms and during periods when raw water is obtained from the lower level of the reservoir that serves as the water source.

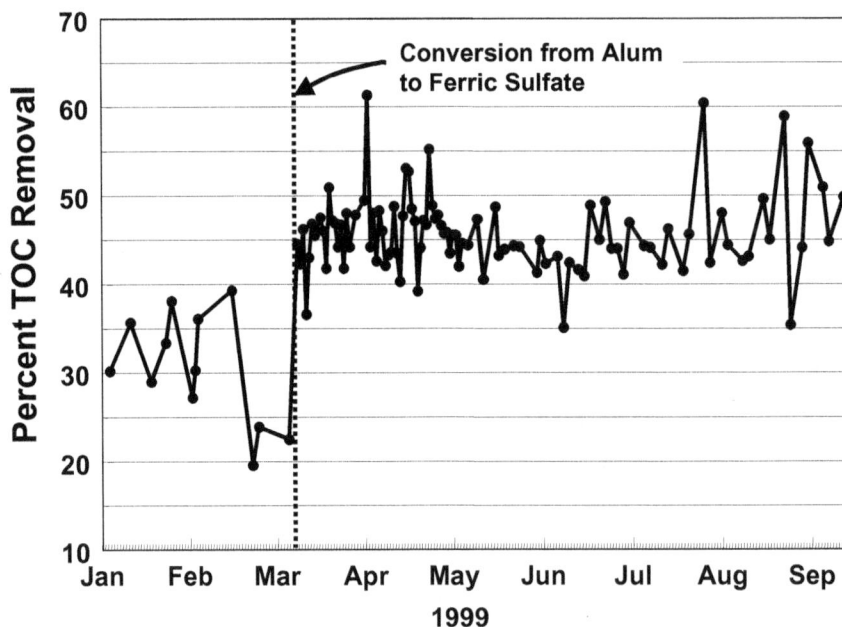

Source: Budd et al. 2004.

Figure 7.1-1 Effect of conversion from alum to ferric sulfate on TOC removal at Chesterfield County, Va.

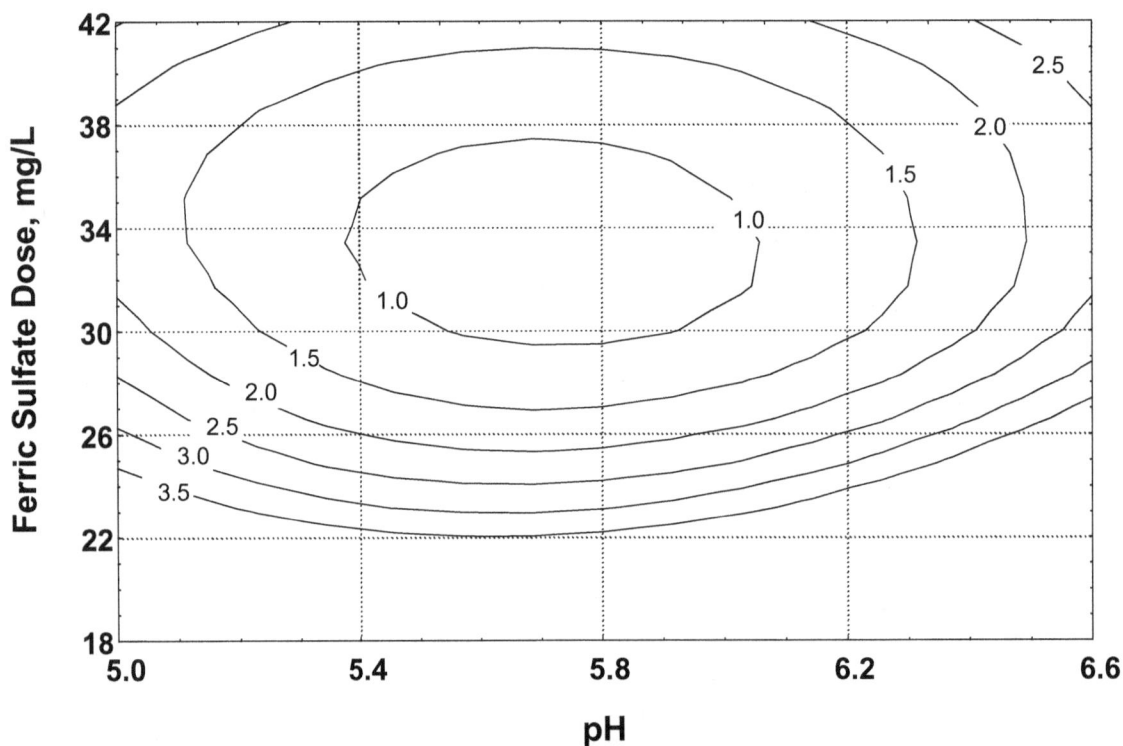

Source: Budd et al. 2004.

Figure 7.1-2 Settled turbidity contours for ferric sulfate at Chesterfield County, Va.

The conversion to ferric sulfate and subsequent improved TOC removal have led to reduced formation of trihalomethanes (THMs) and sum of five haloacetic acids (HAA5) (Figures 7.1-3 and 7.1-4). Hurricanes and periods of bottom withdrawal from the raw water reservoir (identified in these figures) represent special events of heightened disinfection by-product formation; reductions in DBP formation have been observed with ferric sulfate under these conditions as well.

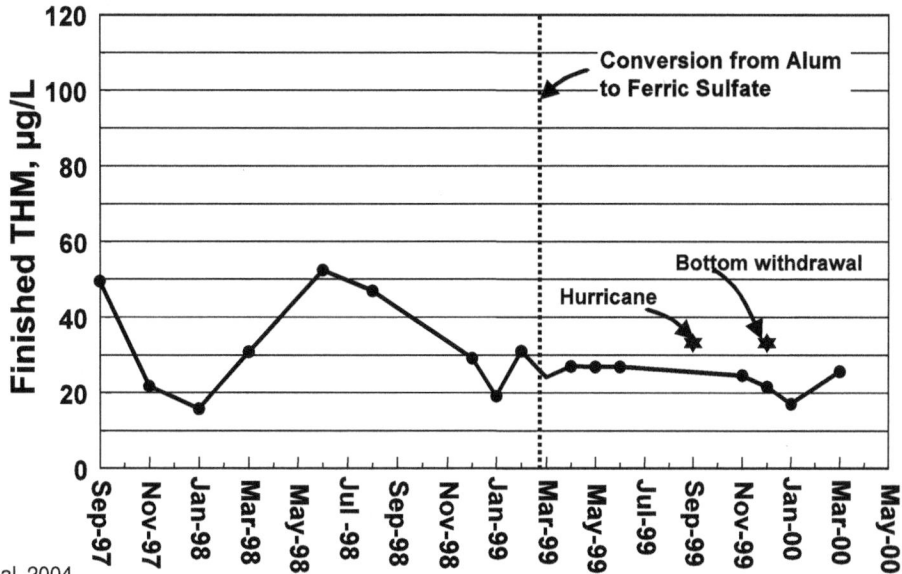

Source: Budd et al. 2004.

Figure 7.1-3   Effect of conversion from alum to ferric sulfate on THM formation at Chesterfield County, Va.

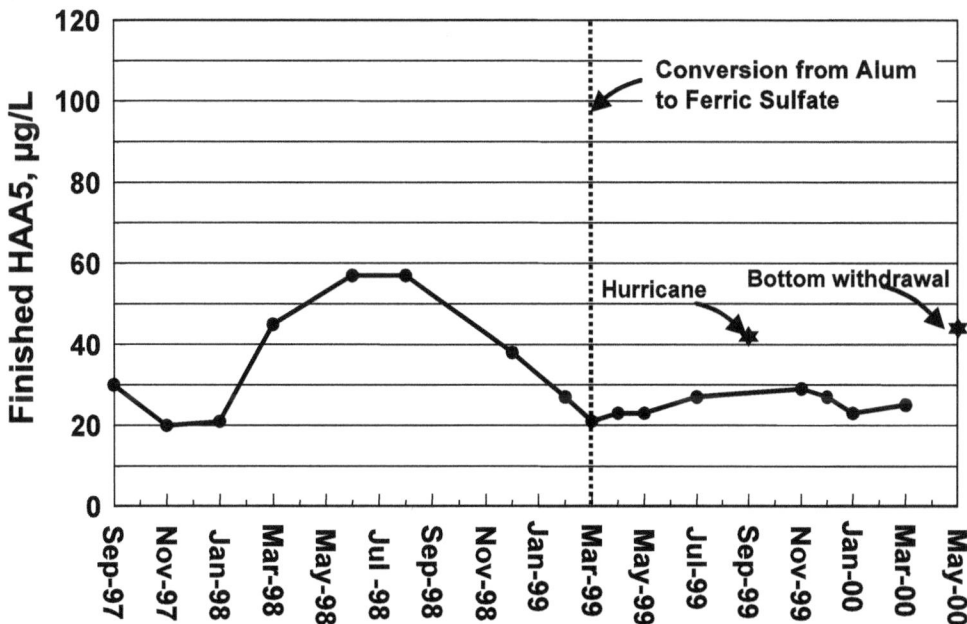

Source: Budd et al. 2004.

Figure 7.1-4   Effect of conversion from alum to ferric sulfate on HAA5 formation at Chesterfield County, Va.

REFERENCE

Budd, G.C., A.H. Hess, H. Shorney-Darby, J. J. Neemann, C.M. Spencer, J.D. Bellamy, and P.H. Hargette. 2004. Coagulation Applications for New Treatment Goals. *Jour. AWWA*, 96(2):102–113.

$$Case \; Study \; 2$$

Jar Test Calibration

George Budd and Paul Hargette

Bench-scale testing (jar testing) is frequently conducted at treatment facilities to simulate coagulation conditions, often with a goal of determining optimum conditions for turbidity reduction. One problem that can occur with jar testing is that some jar test protocols are not representative of conditions in full-scale plants and are not adequate for assessing the effects that coagulation adjustments might have on full-scale turbidity reduction. When conducting jar testing, a key first step is a calibration of jar-test conditions to the full-scale plant to provide a match of settled turbidity at settling times that reflect actual hydraulic loading within the plant. Procedures for using jar testing to provide this type of simulation have been described in this manual and in the literature (Hudson 1981).

An important aspect of the calibration process is the application of rapid-mix and flocculation times similar to those used at full-scale. Mixing speeds of the test apparatus are then adjusted until a reasonable match in results is achieved. Flocculation speeds are tapered as appropriate to approximate the sequence in the full-scale facility. Variation in floc formation with scale has been observed by other researchers (Clark et al. 1994). These types of differences that occur with scale of flocculation may be reflected in jar test results since mix conditions that provide for calibration at the jar-test level of evaluation often produce a lower G value than is applied at full-scale, so mixing speeds may need to be adjusted accordingly.

Appropriate jar test settling times are developed based on simulating hydraulic loading rates in the full-scale clarification basins. A hydraulic loading rate for conventional sedimentation basins is often in the range of 0.5 gpm/ft^2 (1.2 m/h), which equates to a settling velocity of 20 mm/min (0.8 in./min). If a standard 2-L jar-test beaker is used in testing, the distance from the water surface to the sample port is 100 mm (4 in.). Given this distance, solids settling at a velocity of 20 mm/min (0.8 in./min) should be able to sink beyond the sample port within 5 min. Test protocols that allow settling times much in excess of this time will not reflect the settling conditions at most full-scale facilities, and marginal floc conditions that appear to be acceptable at significantly longer settling times will not be effective under actual full-scale operating conditions.

Table 7.2-1 Example calibration testing sequence

Calibration Steps	Plant	Jar Test Identification Number						
		1	2	3	4	5	6	7
Rapid Mix								
Speed, rpm	—	100	100	150	150	150	150	150
Duration, sec	—	180	180	180	180	180	180	180
Flocculation								
Speed (Stage 1), rpm	—	40	25	40	35	35	35	35
Speed (Stage 2), rpm	—	20	12	20	15	20	20	20
Speed (Stage 3), rpm	—	—	—	—	—	10	10	10
Duration/Stage, min	—	20	20	20	20	13	13	13
Settling Time, min	—	7	7	7	7	7	7	7
Alum Dosage, mg/L	—	23	22	23	23	23	23	18
Chlorine Dosage, mg/L	—	—	—	—	—	—	4	4
Raw Turbidity, ntu	—	32	46	32	22	20	20	20
Settled pH	6.6	6.7	6.6	6.6	6.7	6.7	6.9	6.7
Settled Turbidity, ntu	2.0–2.5	7.1	13	6.4	3.8	3.6	1.4	2.5

Source: Budd et al. 2004.

Table 7.2-1 shows an example calibration testing sequence as presented by Budd et al. (2004). The calibration in these evaluations is based on achieving settled water turbidity in the jar test similar to that achieved in the full-scale facility. In this case, it was important to incorporate a simulation of the effect of prechlorination before an effective calibration was attained, as shown in the data for Jar Test 7 in the far right column.

REFERENCES

Budd, G.C., A.F. Hess, H. Shorney-Darby, J.J. Neemann, C.M. Spencer, J.D. Bellamy, and P.H. Hargette. 2004. Coagulation Applications for New Treatment Goals. *Jour. AWWA,* 96(2):102.

Clark, M.M., R.M. Srivastava, J.S. Land, L.J. McCollum, D. Bailey, J.D. Chris-

tie, and G. Stolarik. 1994. *Selection of Mixing Processes for Coagulation.* Denver, Colo.: Awwa Research Foundation and AWWA.

Hudson, H.E., Jr. 1981. *Water Clarification Processes: Practical Design and Evaluation.* New York: Van Nostrand Reinhold.

Case Study 3

Relationships Between Coagulation Parameters,
Winston-Salem, N.C.

George Budd, Paul Hargette, and Bill Brewer

Bench-scale evaluations were conducted in 2004 at the Winston-Salem/Forsyth County City/County Utilities' Neilson and Thomas Water Treatment Plants (Budd and Hargette 2008) to assess the role of coagulation for meeting existing and future regulatory requirements. Evaluations were performed to assess the effect of alternative coagulants and coagulation conditions on plant performance, as measured by settled turbidity, filterability, zeta potential (ZP), and ultraviolet absorbance at 254 nm (UV_{254}). The source waters for the two water treatment plants include a river source and reservoir source, with both sources characterized by low turbidity and alkalinity and low to moderate organics levels.

Figures 7.3-1 through 7.3-7 show the results of evaluations in the two source waters to examine relationships among coagulant dosage, pH, turbidity, filterability index measurements (determined from the rate of filtering a sample through a 0.45-µm laboratory membrane filter according to the method of Shull [1967]), surface charge properties as indicated by zeta potential measurements, and UV_{254} as a measure of naturally occurring organic matter (NOM). Results shown in Figure 7.3-1 illustrate the effects of coagulant dosage and pH on ZP results in testing with alum under conditions of low total organic carbon (TOC) and turbidity. Changes in coagulant dosage have a significant effect on ZP at lower dosages, while changes in pH became more important as dosage is increased. It is noteworthy that a shift from a negative to positive ZP is not confined to a single condition; rather a family of pH and dosage conditions exists where ZP passes through zero.

Results for settled turbidity and filterability under these test conditions are shown in Figures 7.3-2 and 7.3-3. The lowest levels for these parameters occurred at pH values just to the left of the ZP = 0 curve superimposed on these plots. For settled turbidity, an optimum pH occurs with each alum dosage over the range shown in this plot. Alum dosage is a factor in relation to settling efficiency; lower settled turbidity optima are possible at higher dosages. This is to be expected as settling effectiveness is improved as the volume and mass of floc increases with increasing dose (Stumm and O'Melia 1968). Interestingly, this type of variation is not observed for filterability and ZP, which exhibit no discernible variation under optimum pH conditions as dosage is varied.

The corresponding plant operation was at an alum dosage of 8 mg/L and a coagulation pH of 6.65, a condition that was consistent with proximity to the ZP = 0 condition and a low Filterability Index. Operation under these conditions proved effective in spite of higher settled turbidities, producing good filter runs with filtered turbidities well below 0.1 ntu and providing a stable condition for operation. These results provide evidence that low settled turbidity is not always the sole determinant for coagulant optimization. Particles were well conditioned as indicated by the other parameters in this situation.

Figure 7.3-4 shows ZP results from another set of source water conditions in testing of water where NOM levels were greater (case 2). As compared with results shown in Figure 7.3-1, higher dosages are required to reach a ZP = 0 and no low dosage optimum conditions are available for any of the parameters (ZP, filterability or settled turbidity). Settled turbidity became more consistent as an indicator of effective coagulant conditions in the higher dosage ranges where these optima occurred. In this case, the magnitude of settled turbidity became a good indicator of coagulation optimization.

UV_{254} was utilized to assess relationships between NOM and coagulation in this testing. Plots of UV_{254} and ZP revealed relationships that were nearly linear in nature as illustrated in Figure 7.3-5. Similar linear results were obtained in testing performed with ferric sulfate, ferric chloride, and two commercial polyaluminum chloride formulations in multiseason evaluations in the two water sources and in testing with ferric sulfate in two sources in Durham, N.C., as well. Taken in combination, conditions spanned different seasons and TOC levels that ranged from less than 2 mg/L up to levels in excess of 10 mg/L. These observations are consistent with observation that NOM can comprise much of the negative charge content in natural waters; reactions with these charge components are necessary for effective coagulation to take place.

Significant differences occurred in the magnitude of UV_{254} response to changes in ZP between the two sources across cold and warm water seasons. The four plots shown in Figure 7.3-6 illustrate a succession of slopes (b) for coagulation with alum that suggests limited effect of UV_{254} on ZP at one end of the spectrum (b = –0.00014), with progressively stronger relation between the two parameters indicated by increase in slope for the other conditions. Positions of ZP = 0 curves in Figure 7.3-7 follow a trend corresponding to these variations in slope. The flattest slope (–0.00014) indicates little interaction between UV_{254} and changes in zeta potential, providing an example where NOM influence on coagulation is low. This occurred during a low flow winter condition on the river. The ZP = 0 curve for this condition extends to lower coagulant doses and higher pH levels than are observed in the other curves in this figure. A systematic shift in the ZP = 0 curves to increasingly acidic levels and higher coagulant doses occurs as the magnitude of the corresponding slopes increases for the summer condition on this source and for both winter and summer conditions on the reservoir source.

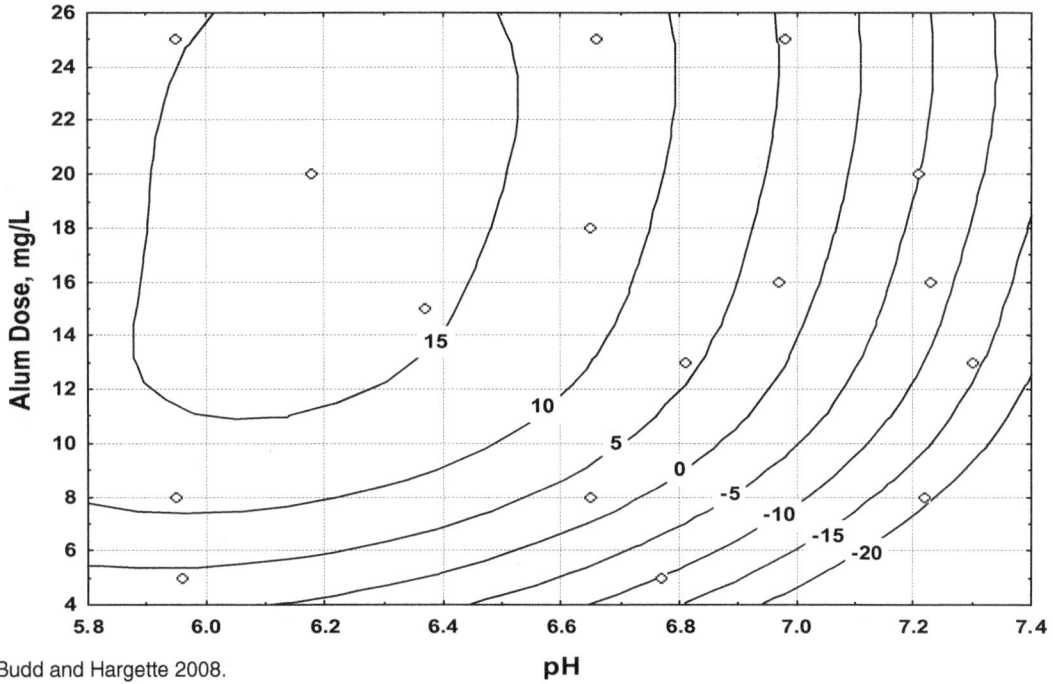

Source: Budd and Hargette 2008.

Figure 7.3-1   Effect of pH and alum dose on zeta potential contours (case 1)

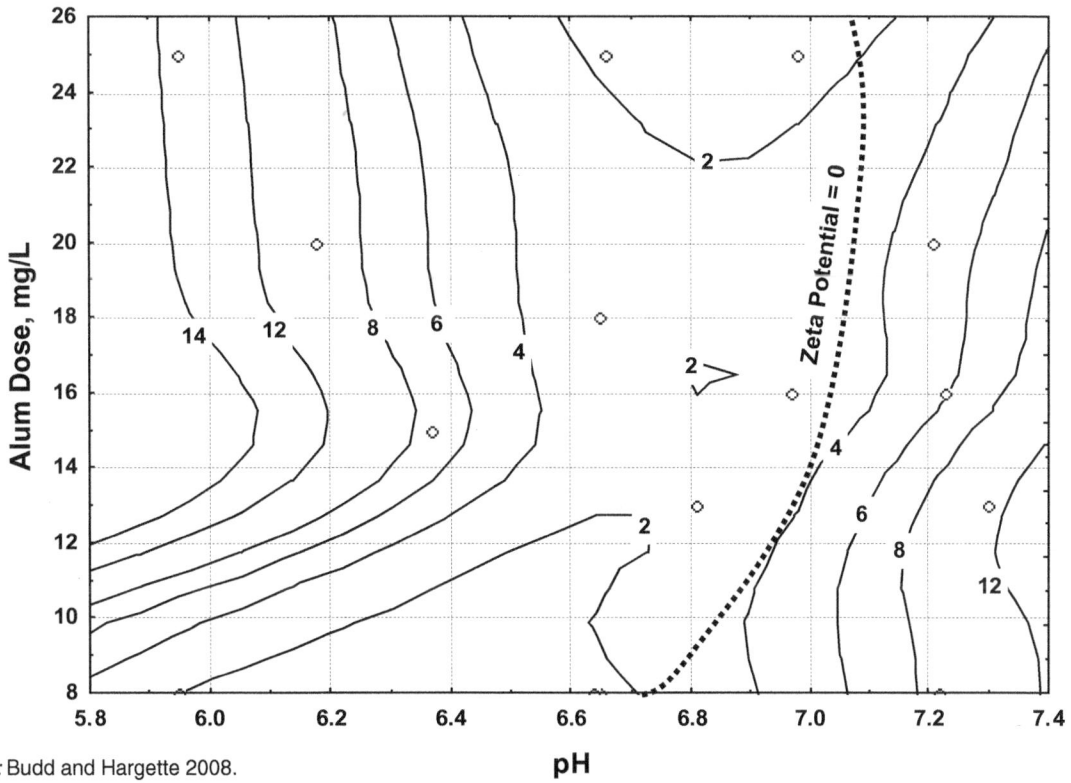

Source: Budd and Hargette 2008.

Figure 7.3-2   Effect of pH and alum dose on Filter Index contours (case 1)

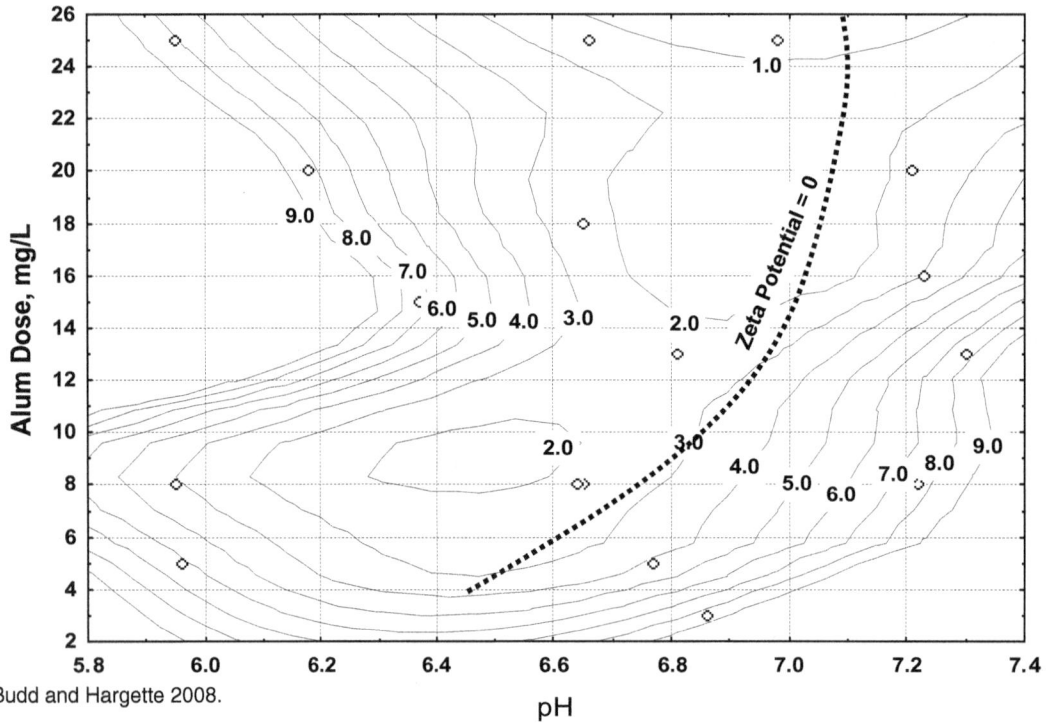

Source: Budd and Hargette 2008.

Figure 7.3-3   Effect of pH and alum dose on settled turbidity (case 1)

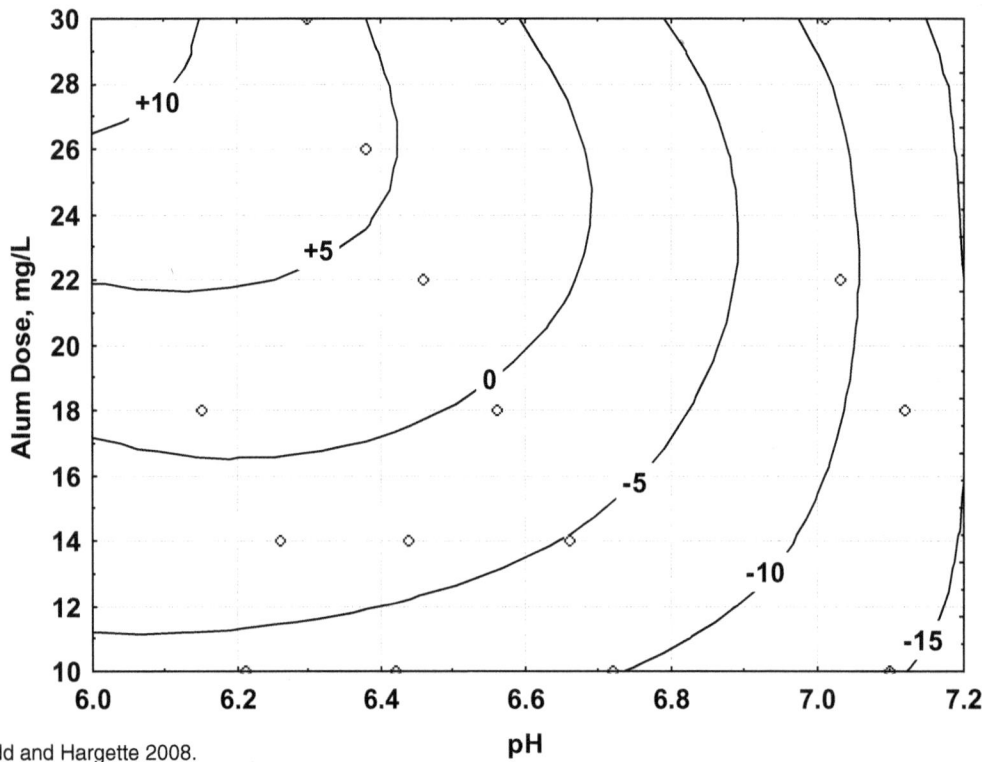

Source: Budd and Hargette 2008.

Figure 7.3-4   Effect of pH and alum dose on zeta potential contours (case 2)

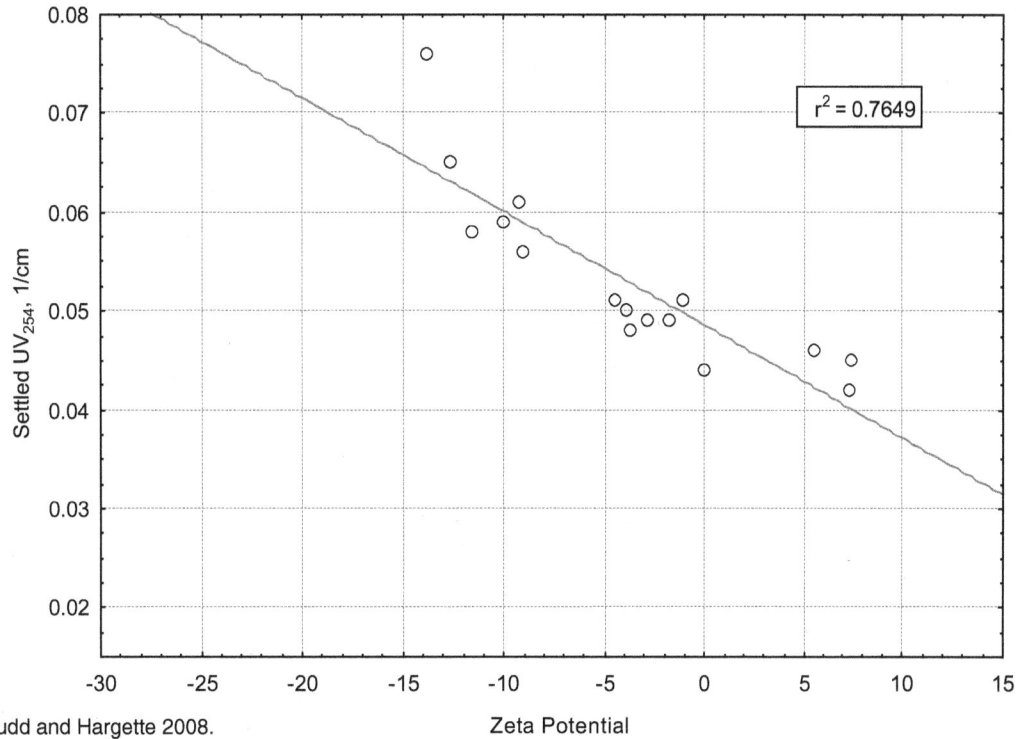

Source: Budd and Hargette 2008.

Figure 7.3-5   Relation between UV_{254} and zeta potential (case 2)

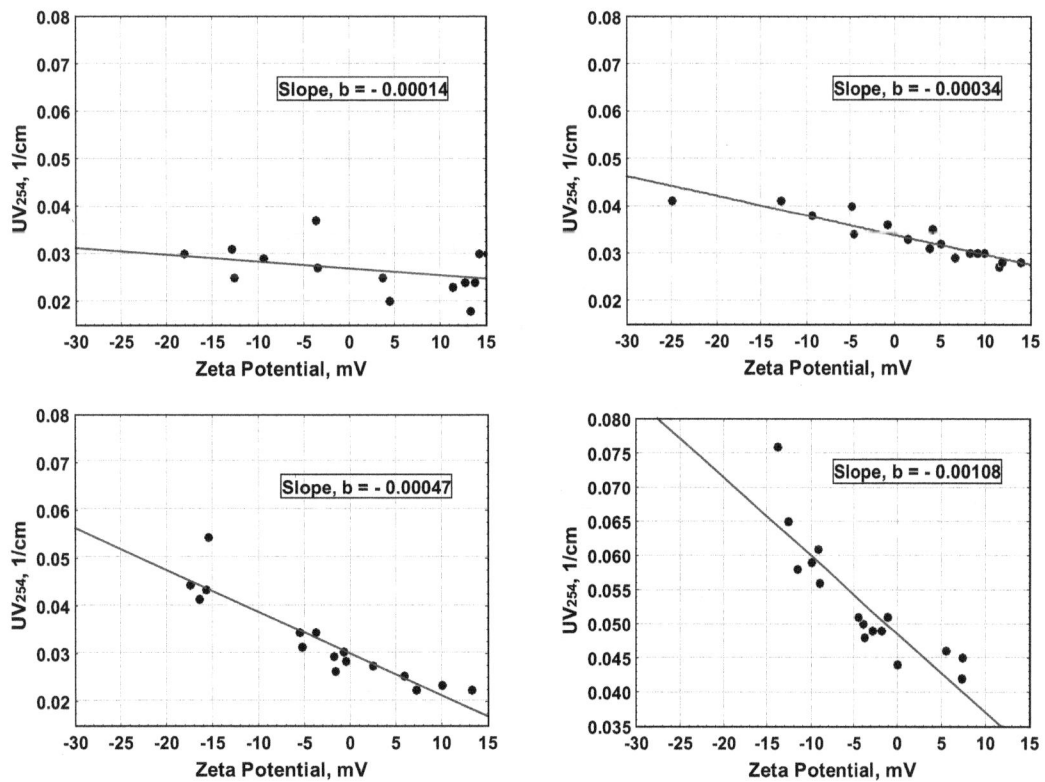

Source: Budd and Hargette 2008.

Figure 7.3-6 UV_{254}–ZP data from multisource, multiseason testing at Winston-Salem, N.C.

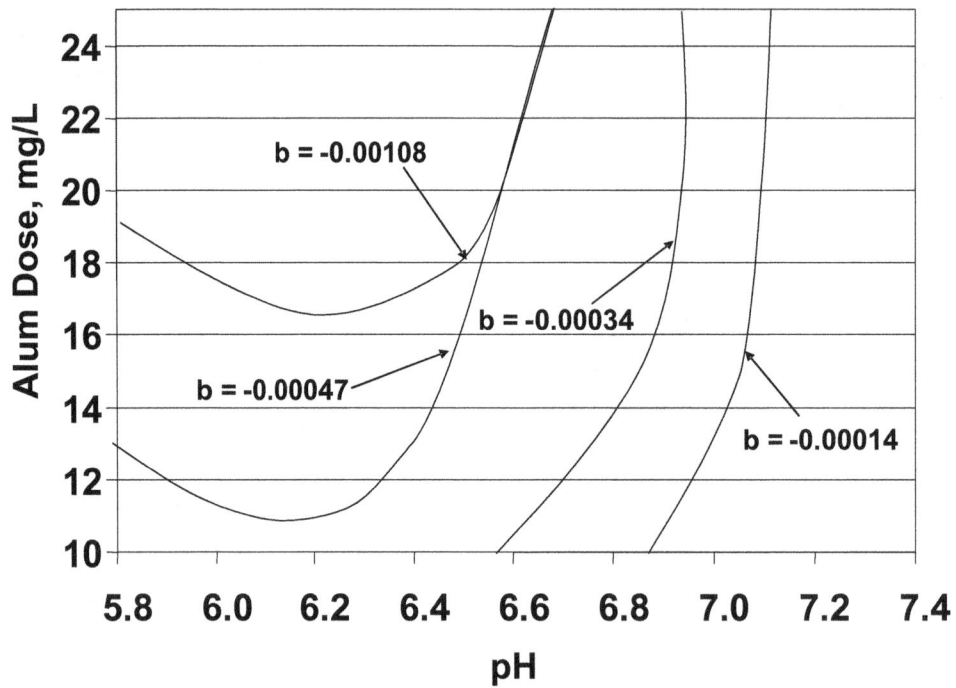

Source: Budd and Hargette 2008.

Figure 7.3-7   Shift in zero zeta potential curves in relation to ZP–UV$_{254}$ slope

REFERENCES

Budd, G., and P. Hargette. 2008. Relationships Between Turbidity, Particle Charge, Filterability, and NOM in Coagulation Process Control. In *Proc. Water Quality Technology Conference.* Denver, Colo.: AWWA.

Shull, K.E. 1967. Filterability Techniques for Improving Water Clarification. Jour. AWWA, 59(9):1164.

Stumm, W., and C.R. O'Melia. 1968. Stoichiometry of Coagulation, Jour. AWWA, 60(5):514.

Case Study 4

NOM Measurements for Coagulation Control

Tom Elford and David J. Pernitsky

BACKGROUND

In drinking water treatment, coagulation is used to destabilize suspended particles and to convert dissolved natural organic matter (NOM) to a solid phase for subsequent removal by solids-separation processes such as clarification, flotation, and granular media filtration. For many raw water sources, the coagulant dose required for proper treatment has been shown to be controlled by the concentration of natural organic matter present in the raw water rather than raw water turbidity. This is true even if NOM removal is not the primary objective of coagulation. This makes measurement of NOM concentration a critical parameter for controlling coagulation processes.

In spite of its importance, many plant operators, especially those with smaller systems, do not have access to current information on raw water NOM concentrations. Depending upon the raw water source, increases in NOM concentration can occur without an increase in turbidity. When this occurs, operators may not be aware of the resulting coagulant underdose until filtered and/or clarified water turbidity begins to increase.

Total organic carbon (TOC) measurements are typically used in the water treatment industry to quantify NOM concentrations. Although high-quality, reliable online and bench-top instruments suitable for water treatment plant (WTP) use are available, their relatively high costs and maintenance requirements mean that many utilities send TOC samples away to off-site commercial labs. The delays inherent with off-site analysis limit the use of this information for making operational decisions.

UV absorbance at 254 nm has also been shown to be an excellent measure of NOM concentration in raw waters, and the analytical procedure is simple, fast, and inexpensive. The exact relationship between UV absorbance and NOM concentration is unique for each raw water source, however. For a given raw water source, increases in UV absorbance indicate increasing NOM concentrations and increasing coagulant demands.

This case study describes the use of UV absorbance and TOC measurements to aid coagulant dose determinations at the city of Calgary Glenmore WTP.

GLENMORE WATER TREATMENT PLANT

The Glenmore WTP is one of two plants serving the city of Calgary, which is located approximately 100 km (60 mi) east of the Rocky Mountains, in Alberta, Canada. The plant is a 350 ML/d (92 mgd) conventional treatment plant treating water from the Glenmore Reservoir, an inline impoundment on the Elbow River.

Turbidity levels are typically low throughout the year, although two runoff events in the spring provide challenging treatment conditions: a lowland runoff and a mountain runoff. The lowland runoff consists of a small turbidity spike but a relatively large increase in TOC as snowmelt carries NOM from forested and agricultural lands. This TOC increase results in a large increase in the coagulant dose at the WTP. The mountain runoff occurs later, once the snowpack in the mountains melts. This runoff event results in high flows and high turbidity because of the erosion of silts in the mountains.

Several bench-top and online instruments are used at the Glenmore WTP to monitor raw water quality, as shown in Table 7.4-1. The plant coagulant dose is selected on the basis of jar tests conducted by plant staff. The dose is verified by monitoring settled and filtered water turbidity. The filters are operated to maintain a filtered water turbidity of less than 0.1 ntu.

Comparison of Raw Water NOM Concentration to Coagulant Dose

To investigate the relationship between raw water quality and coagulant dose, historical plant records were analyzed, and daily raw water turbidity, TOC, and UV absorbance data were plotted against the actual coagulant dose used in the plant for a period of several months in 2003.

Data from a typical spring runoff period are shown in Figure 7.4-1. As described above, a clear increase in raw water TOC and UV absorbance can be seen in April, corresponding to the lowland runoff, followed by an increase in TOC, UV absorbance, and turbidity during the May mountain runoff. Although operators were using changes in raw water turbidity as the primary trigger for increasing coagulant dose at the time these data were collected, the figure shows the close relationship between raw water UV absorbance and TOC and the actual coagulant dose used in the plant. The increased coagulant doses required for proper treatment plant performance (filtered water turbidity was maintained below 0.1 ntu during this entire period) can be seen to be related to the elevated NOM concentrations in the raw water. The April data also illustrate that an increase in coagulant demand can occur in the absence of a large turbidity increase (the peak turbidity during the April runoff period was only 5 ntu).

Table 7.4-1   Analytical instrumentation used at the Glenmore WTP for monitoring raw water quality

Parameter	Instrument Type	Instrument Model
Raw Water UV Absorbance	Bench-top	Hach DR 4000
Raw Water TOC	Online	Hach TOC 1950 plus
Raw Water Turbidity	Online	Hach 1720D (low range) and Hach Surface Scatter 6 (high range)

Courtesy of Tom Elford C.E.T., Glenmore Water Treatment Plant, City of Calgary.

CASE STUDIES 165

Conclusions

From an operational perspective, UV absorbance and TOC measurements have proven to be useful tools for predicting changes in raw water quality at the Glenmore WTP. Increases in raw water UV absorbance and TOC provide operators with an early warning of NOM increases, enabling operations staff to anticipate changes in coagulant dose rather than responding to treated water turbidity excursions after the fact. Furthermore, Glenmore operations staff has observed that when there is an increase in raw water NOM, a slightly higher coagulant dose is needed in the plant to maintain filtered water turbidity than the dose predicted from the settled turbidity results from jar tests. Again, raw water TOC and UV absorbance measurements allow the plant operators to anticipate this situation and to better determine the optimum coagulant dose for the plant.

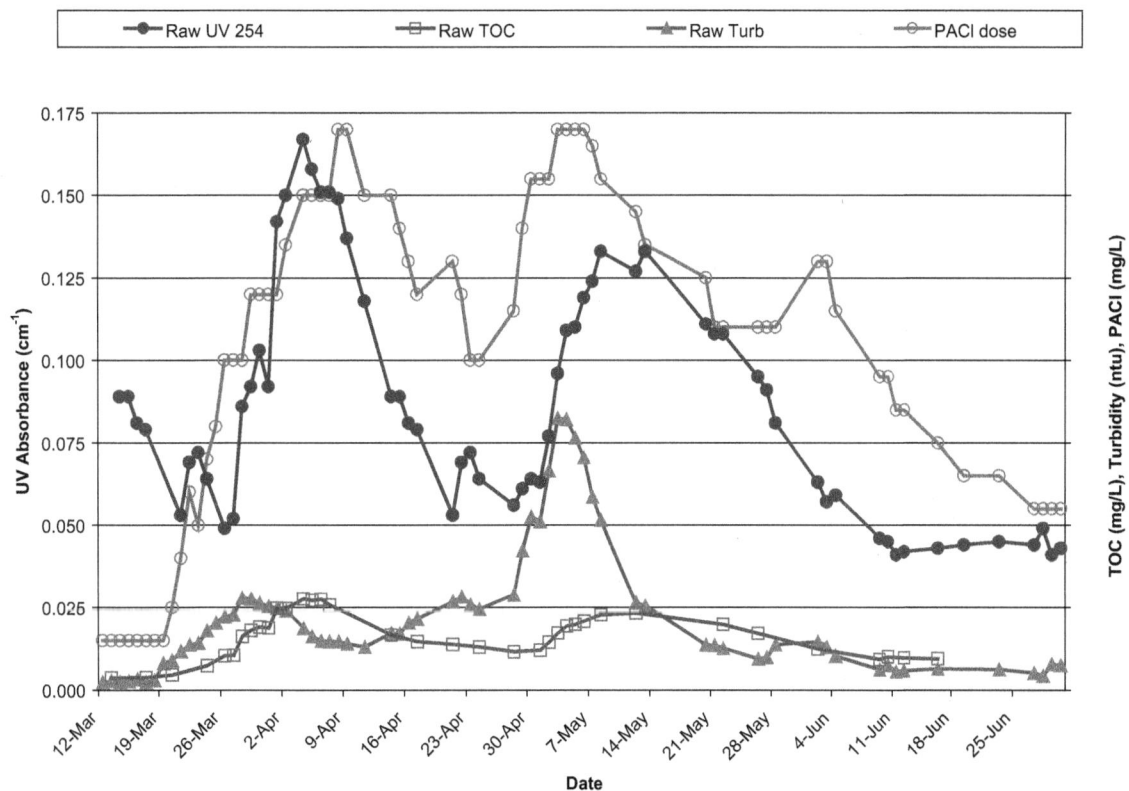

Courtesy of Tom Elford C.E.T., Glenmore Water Treatment Plant, City of Calgary.

Figure 7.4-1   Relationship between raw water quality parameters and coagulant dose for the Glenmore WTP

This page intentionally blank.

Net Charge Equals Positive Change

David Teasdale

In February 2005, Canada's largest low-pressure enhanced coagulation membrane water treatment plant began producing water. Located in Kamloops, B.C., the plant supplies up to 42 mgd (160 ML/d) of drinking water to more than 80,000 residents. Before the plant was completed, pilot studies were done to examine the use of streaming current monitors (SCMs) for optimizing water quality. The results validated the technology's implementation in a distinct and innovative arrangement. Since a streaming current monitor was incorporated in the treatment process in May 2006, a 30 percent increase in membrane performance was noted over the following calendar year. As of December 2008, baseline coagulant dosing has seen a quantifiable reduction of 25 percent overall.

SOURCE WATER

The South Thompson River, the treatment plant's raw water source, is typical of many rivers in the British Columbia interior. It has relatively low alkalinity, low total organic carbon–dissolved organic carbon, fairly consistent pH, with potential for short-term turbidity spikes primarily during events of spring runoff (up to 600 ntu). Like most surface water sources in Canada, the river undergoes substantial temperature fluctuations during the course of a calendar year. Inorganic clay makes up the largest proportion of the river's particulate matter.

For decades, removing fine particles in drinking water has been the primary reason for using coagulants. Traditionally, coagulation has been used to achieve particle destabilization, and most often feed rates are based on source water turbidity levels. As a rule, the higher the turbidity, the higher the coagulant dosage required for effective particle removal and destabilization. At the Kamloops plant, a primary consideration for full-scale operation was to examine whether an online dosing strategy could be developed that would compensate for fluctuating levels of turbidity and natural organic matter (NOM) in the source water—that is, whether it would be possible to develop some sort of online coagulant feed system that automatically compensated for variables that can be seen (turbidity) and variables that cannot be seen (NOM).

Note: Reprinted from Opflow, *July 2007.*

Industry Changes

During the last decade, municipalities treating source waters high in NOM have often needed to use coagulant dosages that exceeded the amounts normally required for particle charge reduction. NOM usually indicates the presence of decaying vegetation in the raw source water. With NOM present, turbidity may have little correlation to effective coagulant dosing, and the percentage of NOM removal often becomes the performance benchmark toward determining adequate coagulant feed rates.

Water treatment plants across North America often use chemical coagulants without a way to quantify proper dosing levels. Other frequently overlooked factors include mixing kinetics and chemical reactions involved in the coagulation process. An online streaming current monitor was incorporated at the Kamloops Centre for Water Quality to help quantify proper coagulant dosing limits.

How It Works

Over the years, instrument advances similar to zeta potential or electrophoretic mobility measurements have emerged in the water treatment industry. The most common is the streaming current monitor, which uses an electric sensor to determine when charge neutralization has been reached in a suspension. The theory is similar to the principle behind a zeta meter in that it is a charge-measuring device whose measurement produces a relative value. The device measures the net ionic and surface charges of colloids in suspension between two electrodes. A piston moves the water back and forth in the chamber, and positive and negative charges are moved downstream to the electrodes where they are measured and generate a streaming current value. The streaming current amplitude and polarity are a function of the sampling location and the type of coagulant used.

Previous piloting work at the plant had identified the "minimum" coagulant feed rate for process optimization in average source water conditions to be 3 mg/L of aluminum chlorohydrate (ACH), the coagulant of choice. That dosage resulted in the best overall membrane performance and optimized organics reduction. Two peristaltic pumps were installed to achieve this optimum dosing level. A flow-based coagulant pump feeds 2 mg/L of ACH into the plant's incoming source water, and a second peristaltic pump runs exclusively from a signal generated by the streaming current monitor and acts in a "top up and trim" fashion. A target streaming current set point is chosen that equates to adding a 1 mg/L dosage of additional coagulant. This system is depicted in the diagram in Figure 7.5-1. The "top up and trim" concept is shown in Figure 7.5-2. Using the aforementioned example, the sum of coagulant, added under ideal water conditions—a key consideration—is 3 mg/L, or the optimum dosing value determined from pilot studies.

This setup allows operators to respond to source water fluctuations in two ways. Adjustments to coagulant feed rates can be made when the streaming current monitor–controlled coagulant pump starts to approach maximum output based on a preset streaming current value. This would result from either a visible change in source water quality, such as a turbidity spike, or an invisible influence, such as an influx of dissolved organics after a heavy rainstorm. The streaming current monitor continuously analyzes changes in source water, while the base coagulant dosing pump maintains a feed rate slightly less than that required for optimum plant performance. Therefore, influxes of organics or invisible influences can be detected and compensated for through an automatic increase in coagulant feed. Without a streaming current monitor in place, invisible influences are more likely to go undetected.

The streaming current amplitude and polarity are a function of the sampling location and the type of coagulant used.

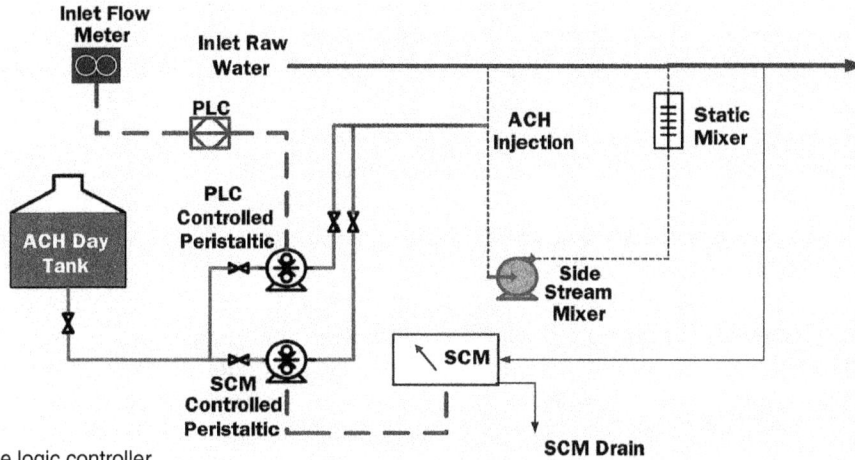

Figure 7.5-1   Kamloops streaming current monitor (SCM) configuration

A target streaming current value set point is chosen that equates to a "top up and trim" dose of 1 mg/L of additional coagulant being added.

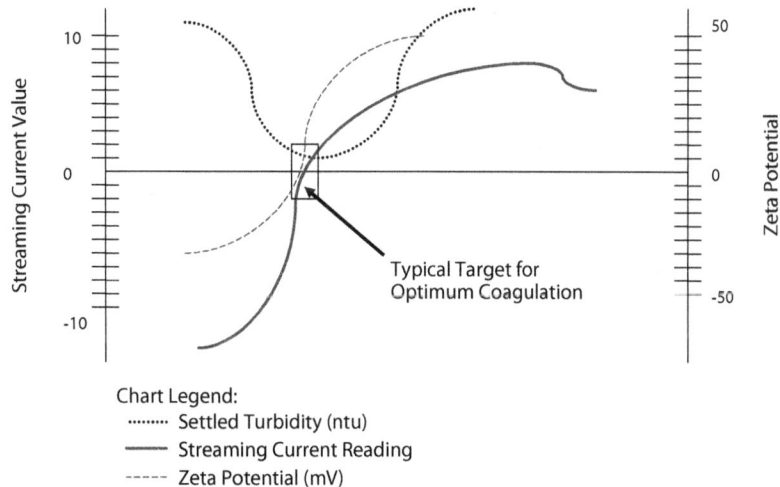

Chart Legend:
········· Settled Turbidity (ntu)
———— Streaming Current Reading
----- Zeta Potential (mV)

Figure 7.5-2   Optimum plant performance

The other obvious benefit is that if either pump should stop operating, there would be a second unit running at all times. If the base-level pump would stop, the streaming current monitor would detect it and automatically compensate its feed rate accordingly. In the event the monitor unit failed, there would always be a minimum base coagulant dosage being fed into the raw water stream—albeit less than ideal. Alarm set-points on the streaming current monitor unit would indicate a problem with the coagulant feed system within minutes. Often in many larger plants, coagulant feed systems can malfunction for several hours before detection.

Unit Optimization

Choosing the "optimum" operating condition for a streaming current monitor can be complicated. The unit produces an on-screen value—usually negative, depending on where the unit is measuring—that corresponds with the level of particulate charge present, the streaming current value. The larger or more positive the value, the more coagulant or positively charged ions are present in the sample. One of the major operational challenges is determining the proper value, or set-point. The objective is to find the optimum streaming current value that corresponds with optimum organics removal, particle destabilization, and plant performance. Streaming current monitor values are not typical of most online instruments, as they must be calibrated to a baseline measurement that will vary from one raw water source to another and from one treatment facility to another. Seasonal water changes may also produce variances in the "target" set-point. Jar tests can help operators determine optimal base-level coagulant dosages when significant changes to source water conditions are encountered.

Other factors that need to be monitored include unit flow rates and piston wear. Both are imperative to the SCM unit operation and the determination of a baseline streaming current value (SCV). When source waters are prone to variable conditions, it may be prudent to consider a unit that is equipped with an automatic flushing mechanism.

Although streaming current monitors are effective and simple in principle, the factors that influence their operation can be complex. This is especially true when operating conditions change or dramatic changes to raw water quality are noted. It is important for operators to understand how such confounding variables may influence the unit's operation, and these limitations should be considered when determining whether a streaming current monitor is appropriate for a particular plant.

Streaming Current Monitor Pilot Study: The Detection of a Ferric Chloride Feed Failure

Michael Sadar

INTRODUCTION

In September 2007, a streaming current monitor (SCM) was placed at a local surface water treatment plant (WTP) for field evaluation. The chief operator at this facility was considering the evaluation of SCM technologies for the monitoring and confirmation of coagulant feed.

The drinking water facility practices conventional treatment for surface water with a production capacity of between 15 mgd (57 ML/d) in winter and 35 mgd (132 ML/d) in summer. Treatment includes chemical injection to adjust pH and the addition of ferric chloride as the primary coagulant. After sedimentation, the water is filtered through dual-media anthracite filtration, followed by disinfection.

The source water originates in the central Rocky Mountains and is primarily from snowmelt. This surface water flows approximately 60 mi (about 100 km) before being captured in an impoundment of approximately 1,000 surface acres (400 ha) that is located in an urban area approximately 3 mi (5 km) from the plant. The reservoir is considered to be a very stable raw water source but is susceptible to flashing from the rare strong thunderstorm.

Three SCM instruments from various manufacturers were installed in the plant to continuously sample the flash mix effluent. SCM signals were logged directly into the plant supervisory control and data acquisition (SCADA) system, but because this was only a trial, the SCM data were visually monitored only and were not wired to any alarms within the SCADA system. The SCM readings were displayed for each of the three instruments, with the values updated each minute. Every 24 hr, the staff would clean each SCM measurement cell and re-zero the instrument. Through the rezeroing of the instrument, the operators can, at a glance, determine if the SCM value is in its expected range. This value is expected to remain close to zero because of the typical stability of the raw water. If the value deviated strongly negative, it is a potential signal that the coagulant feed was lower than the previous optimum, or if it was to deviate positively, the coagulant feed was higher than optimum.

Initial Operation

The first 3 months involved the installation of the sensors, understanding the features of each sensor, and getting used to looking at SCM data as a process performance tool. This was also the time for the plant operators to become familiar with the technology and begin the process of understanding and using the information in their operation practices.

When integrating any new online monitor into plant operations, patience is required. The operators must first understand what the measurement is and how it can be useful in their day-to-day operations. They must understand the critical functions, operational protocols, and maintenance requirements of the instrument to ensure the generated data are reliable. Once the data are proven reliable, operators must learn how different changes in the raw water and chemical dosages in the treatment processes can influence the response of each of these instruments. Having learned this, they can also learn how to use the SCM information to: (1) better monitor their treatment processes; (2) optimize their processes; and (3) troubleshoot unexpected events. Over the first 3 months of the SCM trial, the plant staff learned to understand streaming current technology, its respective strengths and limitations, and the maintenance requirements of the instruments, and how to adjust coagulant doses in response to changes in streaming current output.

Detection of Chemical Feed Failure

Overall, the integration of the SCMs went well. However, it was not until the staff had an unexpected chemical feed failure that was detected by the SCMs, that they fully embraced the use of SCM technology in plant operation.

On February 1, at approximately 1900 hours, an abrupt increase in flash mix pH was seen, as shown in Figure 7.6-1. There was no change in the raw water turbidity or the raw water pH at this time. The operator initially suspected an overfeed of lime. It was not until the operator keyed on the SCM values and examined the SCM trend that he suspected a coagulant feed problem. The coagulant feed pumps were checked, and it was found that one of the coagulant feed pumps had partially lost its siphon, reducing the ferric chloride dose by approximately 50 percent. After the cause was identified and remedied, the operators fully bought into SCM technology as a valuable process-monitoring tool.

The value of streaming current technology can be seen by examining the data in Figures 7.6-1 and 7.6-2 in more detail. In addition to the immediate decrease in the SCM reading after pump failure, Figure 7.6-1 also shows the eventual effect this had on the settled water turbidity. Prior to the partial feed failure, the turbidity of the settled water was about 1.2 ntu. After the coagulant failure, the resulting settled water turbidity increased to 2 ntu. The peak in the settled water turbidity was seen approximately 3 hr after the identification and correction of the coagulant feed failure, emphasizing the early detection provided by the SCM.

A similar trend is shown in Figure 7.6-2 for the filter effluent turbidity. The coagulant feed failure resulted in an eventual increase in the filter effluent turbidity from a base value of 0.04 ntu to a maximum value of 0.09 ntu. This peak value in filter effluent turbidity occurred 3.5 hr after the coagulant underfeed event.

In summary, the underfeed incident caused the finished water turbidity to increase from a minimum of 0.04 ntu to a maximum of 0.09 ntu. The settled water turbidity increased from about 1.3 ntu to 2.3 ntu before returning to baseline. Though the finished water never approached regulatory limits for turbidity, the correlation between the decrease in SCM reading followed by the sequential increase in settled water and then the finished water turbidity is undeniable. The SCMs provided excellent early warning detection but were being used only as pilot study instruments at the time and were not integrated into the SCADA alarm system. If they had been integrated, it is probable that the subtle falloff in the ferric-metered dosage would have been caught and corrected even more quickly with a correspondingly smaller effect on both the settled and finished water turbidity.

Conclusions

The coagulant feed failure event at the WTP was easily detected by the SCM technology. The instrument was very stable and generated a stable baseline. The instrument can be easily rezeroed and this feature provided a consistent recognizable baseline value for the operator. The operators monitor over a hundred parameters concurrently for the operation of the treatment process, so they need a simple reference to determine if the SCM value is good or bad. Being able to set that reference to zero provides a simple means of problem identification, literally at a glance. This feature allows a single-point evaluation of the coagulant dosing system without the need to generate graphs or trend lines.

Following this underfeed event, the operators at this water plant have gained a significant amount of trust and confidence in the SCM instrument and now use it to monitor real-time stability of the chemical dosing system. The plant staff implemented and continues to practice a standard operating procedure (SOP) for cleaning and rezeroing of the SCM instrument on a daily basis. The plant staff also knows that if there is a pH increase combined with an SCM decrease, this indicates a coagulant failure. Without the SCM data, the operator would still have a level of uncertainty and would have to spend additional time troubleshooting the underfeed event. Time is of great importance, as this case has shown, for events upstream can influence downstream processes throughout the entire plant.

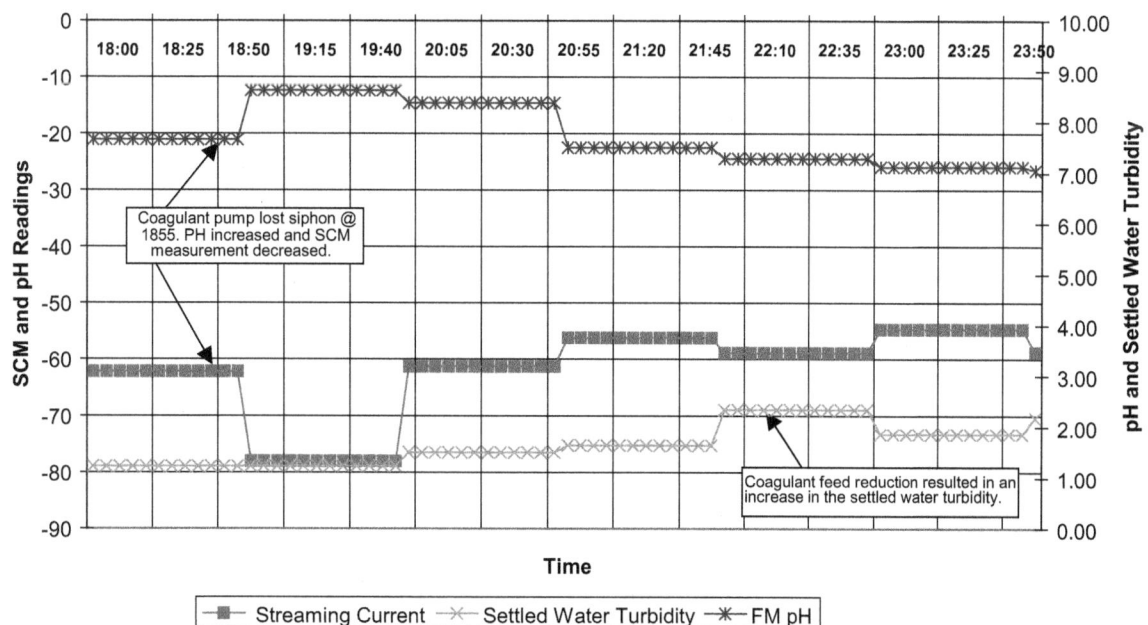

Courtesy of Mike Sadar.

Figure 7.6-1   SCM, pH, and settled water turbidity data during coagulant feed pump failure

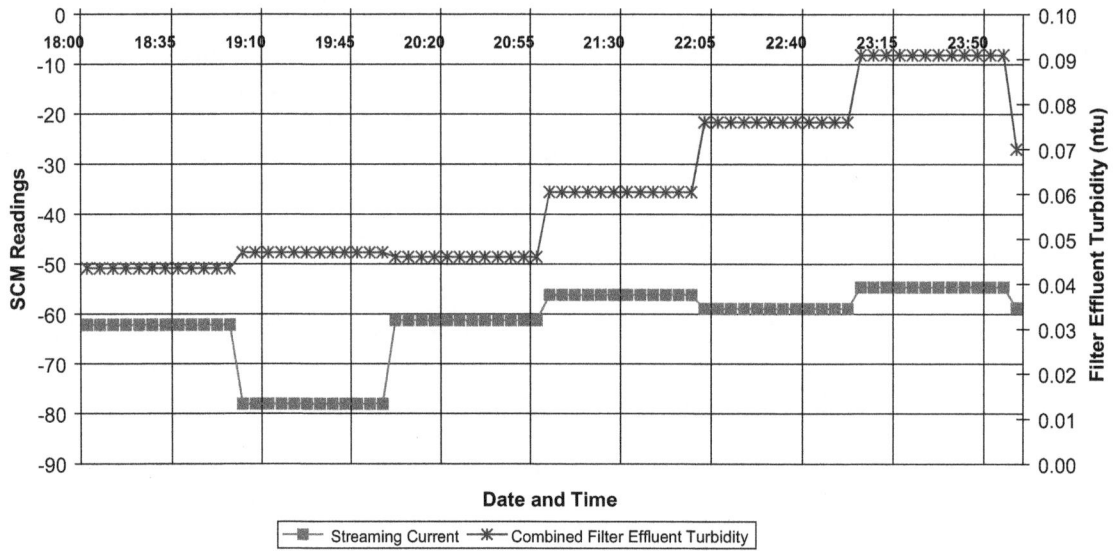

Courtesy of Mike Sadar.

Figure 7.6-2   SCM and filter effluent turbidity data during coagulant feed pump failure

The Application of Simplified Process Statistical Variance
Techniques to Improve the Analysis of Real-Time Filtration
Performance

Michael Sadar

INTRODUCTION

Laser-based particle detection technologies, namely laser turbidity and particle counting, are often utilized to monitor filtration processes. In addition to the raw measurement values that they generate, these technologies can often provide additional information regarding the performance of the filtration process. The baselines of filtrate or permeate product from a filtration system can be more highly characterized, and the variability of the baseline itself can be quantified through the use of simplified statistical techniques.

This case study involves the use of a laser turbidimeter to monitor the filter effluent from dual-media conventional treatment drinking water plant. The purpose of this study was to determine if the laser turbidimeter data could provide any additional information on filtration performance beyond the simple turbidity measurement. Preliminary results indicate that the application of the relative standard deviation (RSD), a simple statistical parameter for laser turbidity or particle counting, can enhance the sensitivity for detecting conditions within a filter that would signify the need for termination of a filtration run.

Drinking Water Plant Background

The drinking water plant involved in this study is a conventional filtration plant that receives raw water from a single surface water source. This source is from a large reservoir, which receives snowmelt runoff from the central Rocky Mountains. The raw water is chemically treated for pH adjustment first, followed by coagulant addition, with the primary coagulant being alum. After the coagulation and flocculation steps, sedimentation is performed through a lamella plate system. This is then followed by filtration and disinfection. During the filtration step, the plant also uses a proprietary filter aid to enhance the effectiveness of the filter run. The plant is a member of the Partnership for Safe Water and has very stringent filtration performance standards, with an internal filter effluent turbidity specification limit that is not to exceed 0.08 ntu. Thus, the plant is in need of filtration monitoring technologies that can accurately detect very small changes in filtration performance at all times.

Test Setup

The laser turbidimeter was designed with heightened sensitivity (in comparision to EPA 180.1 designed instruments) to monitor for the presence of particles in the filter effluent on a randomly selected filter (USEPA 1999, Sadar 1999). The instrument was set up according to the manufacturer's instructions with respect to sampling and to deliver a bubble-free sample for analysis. Signal averaging was set at a 15-sec interval, which was intended to reduce any measurement noise. Data were logged at 30-sec intervals. The laser turbidity values in milli-nephelometric turbidity units (mntu) were also logged at the same interval, so there was no data overlap. After data were collected, a simple statistical parameter was calculated from the particle count and laser turbidity data to provide a measure of the data variability.

The measurement of variability can be quantified through a single statistical process known as the percent relative standard deviation or %RSD (RSD in this study). The RSD is calculated as the standard deviation for a given set of measurements divided by the average for the same set of measurements. This value is then multiplied by 100 to express as a percent. See equation 1 below:

$$\%RSD = (Stdev_n \, / \, Av_n) \, \xi \, 100 \qquad \text{(Eq 1)}$$

where n = a defined number of measurements that are used to calculate both the average and the standard deviation. In the case of this study, n = 7 for both the particle counting and laser turbidity value. Every time a new value was logged, a new %RSD calculation could also be calculated. These values were then logged along with the actual turbidity and particle count values. Thus, for this study, four parameters were in use for monitoring the progression of a filter run: laser turbidity value in mntu, particle counts >2 μm, variability in the turbidity measurement as %RSD, and variability in the particle count measurement as %RSD.

A total of 60 filtration runs were conducted over a period of 90 days. For each filter run, the %RSD was calculated for each parameter and then was plotted alongside the respective turbidity value using the plant's supervisory control and data acquisition (SCADA) system.

Figure 7.7-1 provides a graph of filtration performance for a 24-hr period. During this period, the filter cycle progresses from the ripening phase, at approximately 01:00 hr. The run then progresses until it is terminated by backwash 22:00 hr. When looking at the laser turbidity reading, the values remain stable at about 20 mntu with an average variation of less than 3 mntu through the first 12 hr of the run. (Note that 20 mntu = 0.02 ntu.) After 12:00 hr, the run exhibits increasing variability (%RSD), and individual short-term turbidity spikes increase in both frequency and magnitude as the run progresses. These spikes are most likely a result of low concentrations of particles that are bleeding through the filter as the run progresses. The concentration of particles is not enough to cause a definitive shift in the turbidity value itself, but it does signify that filtration performance is deteriorating over time. Although the variability in the results can be seen in the turbidity data alone, the %RSD calculation magnifies the effects of the variability, making it much easier for operators to see the decrease in filter performance toward the end of the run.

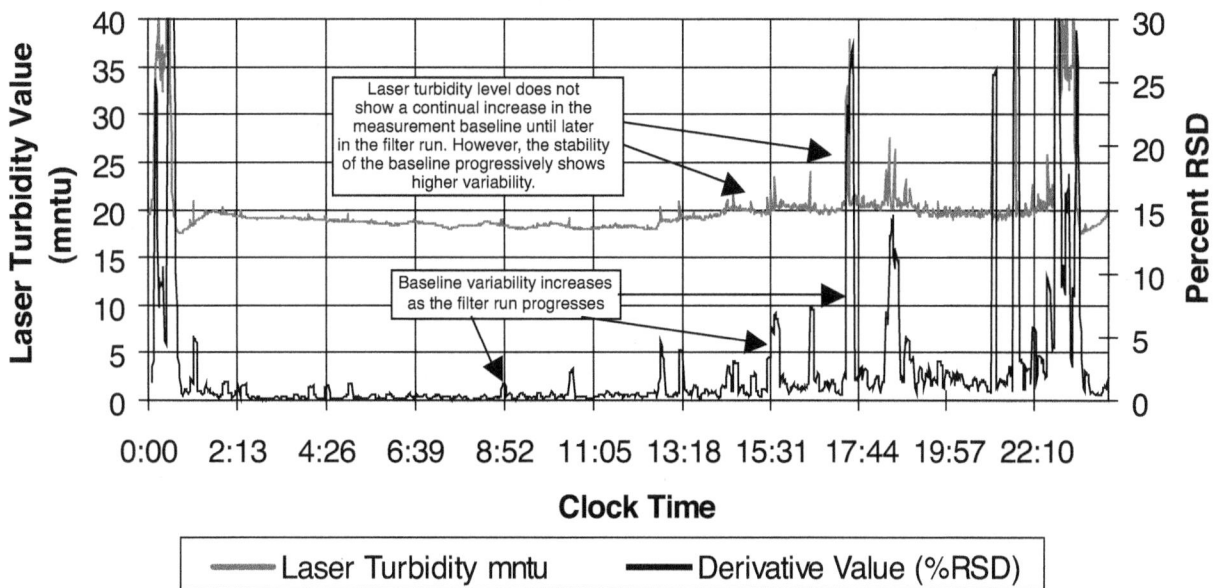

Courtesy of Mike Sadar.

Figure 7.7-1 Measurement of turbidity and the variability of the turbidity measurement from the effluent stream of a granular anthracite dual-media filter

Conclusions

The advent of laser turbidimeters and other particle detection systems has substantially improved the ability to monitor filtration performance. These systems also possess the unique optical qualities of stable and columinated light sources, specific detection angles, and highly sensitive detection systems. The combination of these features provides a very stable process measurement system.

The enhanced stability of the process monitoring system provides additional information regarding the process by monitoring the behavior of the baseline and correlating baseline fluctuation to filtration integrity. This case study provides an example on how fluctuation of the measurement parameter can be used as a separate parameter for monitoring filtration performance, including the deterioration of such performance as a filter run progresses.

REFERENCES

Sadar, M. 1999. *Introduction to Laser Nephelometry: An Alternative to Conventional Particulate Analysis Methods.* Lit. No. 7044. Loveland, Colo.: Hach Company.

United States Environmental Protection Agency (USEPA). 1999. *Method 180.1, Determination of Turbidity by Nephelometry.* Cincinnati, Ohio: USEPA.

This page intentionally blank.

Online Monitoring Aids Operations at Clackamas River Water

Robert D. Cummings

Online monitoring plays a key role in the efforts of operators at Clackamas River Water's (CRW's) C.R. Harrison plant to economically produce filtered water having excellent quality. As described in this case study, online monitoring is used extensively throughout the treatment plant, which was built in 1964 at 10 mgd (38 ML/d) capacity and was expanded to 20 mgd (76 ML/d) and finally to 30 mgd (113 ML/d).

The source of water for this plant is the Clackamas River, which begins on the slopes of Ollalie Butte near Mt. Hood and flows 82 mi (132 km) from an elevation of 6,000 ft (1,800 m) down to 12 ft (2.2 m), where it meets the Willamette River downstream from Clackamas River Water's intake. The watershed drains an area of nearly 939 mi^2 (about 2,420 km^2), winding through forests, mountain meadows, farmlands, a light industrial area, and suburban neighborhoods. More than 200,000 Oregonians depend on the Clackamas River for their supply of high-quality drinking water, hydroelectric power, and outdoor recreation. The watershed supports a rich variety of native plants, animal species, and their habitats.

The Clackamas River flow is regulated to some extent by dams, but during fall through spring the full stream flow resulting from rainstorm events is passed through. The turbidity ranges from a low of about 1 ntu in the summer to as high as 900 ntu during a major storm event. Winter storms typically result in 10 to 300 ntu raw water. Pacific storm systems can cause river turbidities to rise at a rate in excess of 0.5 ntu per minute and drop at a similar rate. The raw water pH ranges from 6.3 in winter storms to a high of 8.5 in the summer. The summer pH range may shift from a pH of 7.5 to 8.5 in the same day, because of sunlight changing the algal activity level.

The plant employs a version of direct filtration, with initial mixing provided by a hydraulic jump, followed by two pretreatment trains consisting of rapid mix and contact basins that provide 32 min of theoretical retention time at design flow. Six mixed media filters are made up of 18 in. (46 cm) of anthracite coal, 9 in. (23 cm) of silica sand, and 3 in. (8 cm) of garnet sand. Plant filters are washed at 10 ft (3 m) of head loss, or in case of breakthrough, at 0.10 ntu. Filter runs can be 4–5 hr in extreme river crests to 40 hr in good raw water conditions. Postfilter turbidity annual average is about 0.03 ntu, and finished pH is held close very close to the target of 7.25 units. A combination of automatic and manual controls are adjusted by the water treatment plant operators to take advantage of all the equipment and analyzer precision available.

For coagulation of the source water, aluminum sulfate is added prior to aluminum chlorohydrate (ACH). A hydraulic jump at the ACH application point provides good agitation prior to the two 68-rpm mechanical mixers. ACH performs about 80 percent of the coagulation, being about four times as effective as alum on the source water, and contains about three times the aluminum of alum. The use of ACH as the primary coagulant has eliminated the need for adjusting the alkalinity for coagulation. This has provided a more efficient way to deal with balancing treatment in storm conditions and reduces the overall need for pH-adjustment chemicals. The combination of ACH and alum has been found superior to either chemical alone. Alum continues to be useful for removing small particles and color. The alum and ACH application rates and the total coagulant applied in terms of alum equivalent ppm are tracked by the supervisory control and data acquisition (SCADA) system.

The plant has been given a direct filtration rating because the contact basins have simple baffling at the inlet and outlet, and because of the relatively short (about one half hour) retention time with no provision for high-rate clarification. Filter-aid polymer addition is located at the contact basin overflow weir, and a pipeline parallel to the weir delivers polymer through 86 orifices to aid in mixing.

Chemical feed rates are monitored and recorded with the plant SCADA system. Variable speed/variable stroke pumps are used in most applications, and the pump speed, pump stroke, and resulting ppm concentrations are recorded and archived by the SCADA computers. SCADA software programs allow plotting and scaling of all plant variables. There are three screens available to view SCADA information in the control room, one in the laboratory, one in the operations/maintenance office, and two in the plant manager's office. The screens can display both long-term and short-term water quality and plant performance trends, enabling operators to know what happened on previous shifts as well as the details of recent operations. Observing trends is especially helpful during times when water quality is changing, as it indicates potential for future water quality. Some process variables are displayed on large digital displays and an enunciator panel, as well as the video terminals. A raw water turbidimeter near the river intakes was used as an early indicator of turbidity changes, but the staff opted not to replace a failed unit.

The intake screens have sensors that indicate river level versus postscreen level for screen head loss measurement. Intake screen air-burst cleaning operation system status is monitored by a local alarm enunciator panel, and water treatment plant (WTP) SCADA computers monitor conditions. A level indicator in the river pump station provides for river pump shutdown in the event of screen blockage. Low lift pumps are capable of being throttled to produce prescribed flows or to increase CT values. A voltage monitor reports on each leg of the three-phase power supply to the plant. A tap and conduits for control wires and chemical tubing are in place at the river pump station for any temporary chemical feed that would require more contact time or separate feed locations. Operator-adjustable minimum and maximum river level alarms are available, as are raw water turbidity level and rate of change alarms.

To optimize coagulation, operators at the plant base process coagulant changes on multiple data sources. A streaming current monitor (SCM) is located close to the rapid mixer for fast results. A second SCM, an older but serviceable unit, is continuously used to sample raw water, prefilter water, or other locations as needed. The second SCM usually acts as relative verification of the rapid-mixer SCM. The SCM and raw water turbidity readings are monitored by the SCADA system, and alarms are active for raw water turbidity rate of change, a low SCM limit, high SCM limit, and an operator-adjustable rapid change in SCM alarm.

A dual pilot filter system is used to optimize coagulant application. Each of the mixed-media pilot filter columns is washed each 45 min, filtered-to-waste for 15 min, and then monitored for 30 min. The pilot filter source water is a rapid-mixer sample. No filter aid is used for the pilot filters, unlike the plant filters. The rapid-wash cycle timing and absence of filter aid application keep the pilot filters more sensitive to breakthrough than the plant filters, offering a margin of safety. When turbidity from the pilot filters is less than 0.2 ntu, this is a good predictor of successful full-scale filtration. Cam timers were used for pilot filter valve signal operation, and plant staff intends to continue with this choice for now, though microprocessors are also reliable and offer more adjustment options. A centrifugal grit separator is used to remove sand/silt that can cause the pilot filter valves and SCMs to fail, particularly when they are most needed, in storm conditions. The pilot filter operation is monitored by the SCADA system and employs a high pilot filter turbidity alarm. Pilot filter columns have easy service access, and the air-operated solenoid valves are connected to clear plastic drip tubes to show diaphragm leaks.

Jar testing is performed according to a standardized procedure and usually includes a control jar plus a range of dilutions and additions from the control, or current process conditions. Jars are usually observed visually and may be filtered through Whatman #1 or #2 filter paper. The slightest amount of milky appearance in a lighted jar at the end of the jar test cycle (best with room lighting lowered) reflects poor filterability. Settling rates are visual only, as this is essentially a direct filtration plant. The jar test apparatus has been fitted with an employee-made filter rack attachment that makes sample filtration more efficient.

Jar testing and full-scale pilot testing of organic polymers have always included use of some alum; polymer suppliers and staff have consistently recognized this need. Preblended combinations of polymer and alum or aluminum chlorohydrate (ACH) have not been favored because of the lack of operational flexibility.

CRW uses an online pH meter to augment operator grab samples of rapid-mixer water. Optimum coagulation/flocculation is often found in a pH range of 6.2 to 7.2, depending on specific conditions. Treatment of high color levels have required treatment at or below pH 6.2, requiring specific pH maintenance that is aided by the online monitor. The alkalinity of the Clackamas River ranges from 13 to 32 mg/L as $CaCO_3$ from winter to summer respectively, and even small changes in coagulant may have a measured effect on treatment. Online pH monitors are used on the raw water, following coagulant application, pH adjustment, and finished water pH. High and low alarms are present.

Chlorine residual is monitored online near the point of prechlorine application, prefiltration, after postchlorination, and at the finished water stage. The feed rate in Cl_2 lb/day is monitored and recorded. Prechlorine residual, postfilter chlorine residual, and finished water chlorine residual (high and low) alarms are present to assist operators.

Membrane probe-type chlorine analyzers are used for flash mixer sampling, where fouling of DPD (N,N-diethyl-p-phenylenediamine) colorimeters proved to be a problem. The accuracies of membrane probes are acceptable, and reagent costs were reduced. A centrifugal grit separator is also used to remove sand and silt that can cause prechlorine analyzer failures.

The online monitoring data gathered plus flow data are used to calculate CT values on a continuous basis. CT parameters are continuously monitored and displayed in terms of CT required, CT actual, CT ratio (actual to required), and CT ratio average. Alarms for low CT ratio and low CT are included. Key CT parameters may be displayed on a dedicated monitor connected to the SCADA server.

Filter aid polymer is blended automatically, and blender operation alarms are displayed by the SCADA system; filter aid polymer concentration is also displayed.

Chemicals are applied based on a programmable logic controller (PLC)–based compound loop control, with raw water flow and process instrument or operator input being the two variables. Soda ash for pH adjustment is an exception, where filter flow was chosen over raw water flow for more accurate proportions. Operators locally and remotely monitor chemical pump speed and stroke. Historical chemical feed information is logged on SCADA computers for review and planning purposes.

The six plant filters are controlled from three consoles overlooking the filter beds. Filters are run in semiautomatic mode, with operators initiating an automatic backwash sequence; but filters can also be washed on programmed set-points. Operators may change the backwash flow set-point during automatic backwash operation or on a prescribed visual assessment period, extend the backwash time in 30-sec increments. To extend the backwash of a filter, the operator may press a "backwash extend" button once for each 30-sec extension desired. Periodic changes in the automatic backwash set-point make this sort of adjustment rare. Fully manual operation of plant filters is provided, and pump/valve controls include glove switches, push buttons, and dials (potentiometers). An abort button is available if there is a problem with the automatic backwash sequence. A head loss–based "backwash required" warning alarm is used to prompt filter backwash as needed. Dimmable lighted (LED) bar graphs display filter head loss, filter effluent flow, filter effluent flow set-point, backwash pressure, backwash flow, and backwash flow set-point. Filter status indicator lights include in service, stand by, auto ready, backwash required, backwashing, PLC failure, backwash sequence failure, extend backwash, abort backwash, valve mode/position indicators, and pump operation indicators.

Each filter effluent valve and the common master backwash valve have independent program integral derivative (PID) controllers. Plant filters have individual level indicators that are balanced to a level indicator in the filter influent flume to position the filter effluent valves.

Filter effluent turbidity alarms are set for 0.10 ntu, and with the use of 10- to 15-min filter-to-waste as part of the backwash sequence, they are seldom activated. Other indicators include alarms for water pressure, pump failure, water flows, communication failures, programmable logic controller failure, and so on.

The extensive use of online instrumentation has enabled operators at the C.R. Harrison plant to quickly evaluate plant operating conditions and the quality of raw and treated water. The abundance of up-to-date information provides a basis for making sound decisions about treatment.

Palm Beach County Water Utilities Water Treatment Plant 8 Ferric Chloride Addition

Tim McAleer and Jose Gonzalez

BACKGROUND

The Palm Beach County Water Utilities Department services over a half million residents in southeast Florida with six water treatment plants, of which two utilize lime softening processes. The other four plants are membrane treatment facilities. Five of the water plants are connected to a single distribution system, and the Lake Region Water Treatment Plant located in Belle Glade, Fla., services the cities of Belle Glade, Pahokee, and South Bay.

Water Treatment Plant (WTP) 8 is located in the western portion of West Palm Beach and supplies water to the northern portion of the Palm Beach County service area of approximately 36,000 connections and a population of approximately 91,370 customers.

The treatment processes at Water Treatment Plant 8 consists of 20 mgd (76 ML/d) of lime softening by Eimco treatment units, followed by ozone and filtration. The plant also uses 10-mgd (38 ML/d) of lime softening by an Infilco Accelator treatment unit followed by filtration and anion exchange treatment for total organic carbon (TOC) and color removal. These processes make up the total 30-mgd (114-ML/d) capacity for the plant. Both treatment processes use sodium hypochlorite and ammonia for disinfection purposes.

Polymer Testing

During the period between 2000 and 2005, many different polymers were being tested at WTP 8 because of the occurrence of turbidities over 7.0 ntu after the softening process, which caused sludge settling inside of the ozone contactor and shortened filter runs. The sludge from the softening process was quite dense and settled too fast to be recirculated in the treatment units, so it accumulated on the floor of the clarifier and eventually caused the rake system to shut down for high torque. This required a shutdown of the clarifier to remove the water and hose out the sludge.

Maintenance

This heavy sludge being produced from the high-molecular-weight polymer did improve the effluent turbidity from the clarifiers but was causing problems with the ability to remove the sludge from the clarifiers and in the sludge thickeners used with the vacuum filter sludge drying system. The sludge thickeners also used a rake mechanism to move the sludge toward the pump to send it to the vacuum filter, but the sludge was causing excessive torque on these rake mechanisms just as it did in the clarifiers.

The plant operators were backwashing filters every 75 hr to keep the filtration system working well without having problems with differential pressure on the filters. The backwash water was left to settle in a basin before being pumped back to the raw flow pipes on each of the three clarifiers, but the sludge from the filter backwashes was too heavy to be pumped to the sludge thickening process for thickening and subsequent drying as the system is designed. This required the operations and maintenance staff to take the backwash recovery basin offline for 1 week each quarter to remove the settled sludge from the basin.

These maintenance issues created a backlog of work caused by the shutdown of plant processes at any moment for a cleaning procedure.

Water Quality

The finished water quality at WTP 8 has always been well within established regulatory standards, but there was room for improvement in the areas of color, turbidity, and TOC removal in the finished water. Palm Beach County Water Utilities had made the decision to expand the plant's capacity with 10 mgd (38 ML/d) of anion exchange process to lower the color and TOC in the finished water.

The raw water quality at WTP 8 is very consistent as shallow aquifer wells are used for the treatment process. Table 7.9-1 shows the raw water quality along with the finished water quality before the addition of ferric chloride to the treatment in September 2006.

Customer Complaints/Inquiries

Table 7.9-2 shows water quality inquiries from 2004 to 2008. The polymer testing was taking place during 2005 and most likely had much to do with the higher-than-normal customer inquiries during that year. After the ferric chloride addition began in September 2006, there was a noticeable reduction in customer water quality inquiries in the following years.

In 2006, Palm Beach County Water Utilities decided to test ferric chloride addition to their treatment process after learning that Delray Beach Water Treatment Plant had used that chemical with excellent results in their Eimco treatment units

Table 7.9-1   Water quality characteristics before ferric addition

Analyte	Raw Water	Finished Water (Before Ferric)
TOC	10–13 mg/L	7–8 mg/L
Color	25–30 units	7–9 units
pH	7.2	9.0–9.2
Total Hardness	260 mg/L	70–80 mg/L
Turbidity	0.10–0.20 ntu	0.10–0.30 ntu

Source: Palm Beach County.

Table 7.9-2   Customer complaints showing reduction after 2005

2004	2005	2006	2007	2008
57	100	44	45	46

Source: Palm Beach County.

Table 7.9-3   Water quality characteristics after ferric addition

Analyte	Raw Water	Finished Water (After Ferric)
TOC	10–13 mg/L	2–4 mg/L
Color	25–30 units	1–3 units
pH	7.2	9.0–9.2
Total Hardness	260 mg/L	70–80 mg/L
Turbidity	0.10–0.20 ntu	0.01–0.05 ntu

Source: Palm Beach County.

after experiencing the same issues with thick sludge and excessive carryover from the clarifiers onto the filters. WTP 8 staff performed jar testing using ferric chloride at dosages of 10 mg/L to 15 mg/L to evaluate the effects of the chemical on settling in the clarifiers at differing dosages. The plant operations staff decided that they would start at 12.0 mg/L and evaluate the process again to determine the need for increasing or decreasing the dosage.

Palm Beach County contracted with a rental tank supplier and PVS Technologies to supply the ferric chloride chemical. The plant personnel ordered metering pumps approved for ferric chloride and installed piping to the clarifiers.

On September 9, 2006, ferric chloride addition was started on the two Eimco treatment clarifiers that were in service at the time while the plant anion exchange expansion was under construction. The finished water quality was improved by the use of ferric chloride in the terms of color, TOC, and most importantly turbidity, as shown in Table 7.9-3. The addition of ferric chloride also reduced the amount of water used for flushing in the distribution system for water quality complaints.

Within 2 hr of starting the ferric chloride addition to the treatment process in the clarifiers, there was a very noticeable decrease in carryover from the treatment units, and the operations staff were able to increase the amount of recirculated sludge in the clarifiers without any excess torque showing on the rake system. The sludge from the process was much easier for the rake system to move to the center of the clarifier to be recirculated and flowed freely through the blowoff system to the sludge thickening process. This allowed the rake in the sludge thickener to move the sludge to the center of the basin, where it could be pumped into the vacuum filter process for dewatering and subsequent drying.

Sludge Analysis

There was some concern in the early stages of ferric chloride addition as to whether the plant vacuum filter would be able to dry the sludge sufficiently with the ferric chloride, because that chemical causes the sludge to retain the water and can make dewatering

Table 7.9-4 Sludge analysis for land application purposes

Area of Sludge Pile Sampled	Arsenic mg/kg	Manganese mg/kg	Iron mg/kg	Aluminum mg/kg	Copper mg/kg
Top	No detection	0.00320	4,670	No detection	3.37
Middle	No detection	No detection	4,150	No detection	3.51
Bottom	No detection	0.00330	3,460	0.0634	1.84

Source: Palm Beach County.

more difficult at high dosages. There was also some concern as to the content of iron, manganese, copper, aluminum, and arsenic in the sludge after the process and whether it would still be suitable for land application after adding the ferric chloride chemical. Table 7.9-4 shows the levels of these contaminants found in the sludge after the drying process.

Figures 7.9-1 and 7.9-2 show the turbidity reduction from August 2006 to September 2006, after the ferric chloride system was put online. The carryover from the lime softening clarifiers was drastically reduced, allowing for longer filter runtimes and fewer customer complaints in the distribution system.

Chlorine and Polymer Reduction

After the ferric chloride system was placed into service with such positive results the staff at WTP 8 decided to lower the polymer dosage on the clarifiers. The polymer dosage was reduced in small increments from the starting point of 0.15 mg/L and ultimately was lowered to a 0.02 mg/L dosage rate with no change in the settling ability of the sludge in the clarifier or an increase of carryover.

The ferric chloride addition helped to lower the TOC in the water leaving the clarifier, and combined with the lower carryover, the plant staff noticed a chance to lower the chlorine dosage with no detrimental effects to the chlorine residual in the distribution system. The cost savings from the chlorine and polymer reduction helped to offset the cost for operating the system.

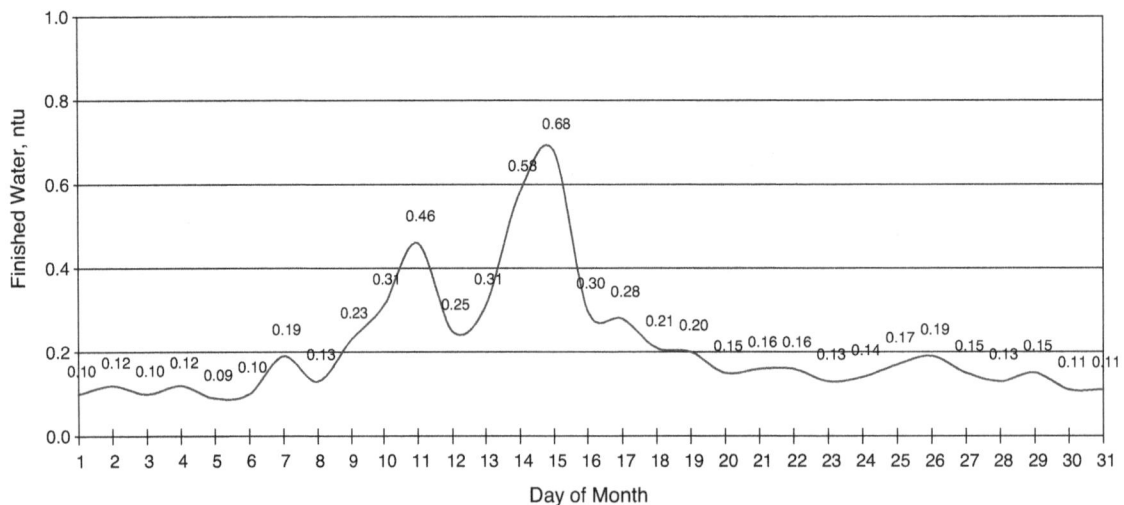

Source: Palm Beach County.

Figure 7.9-1 Turbidity before addition of ferric chloride, August 2006

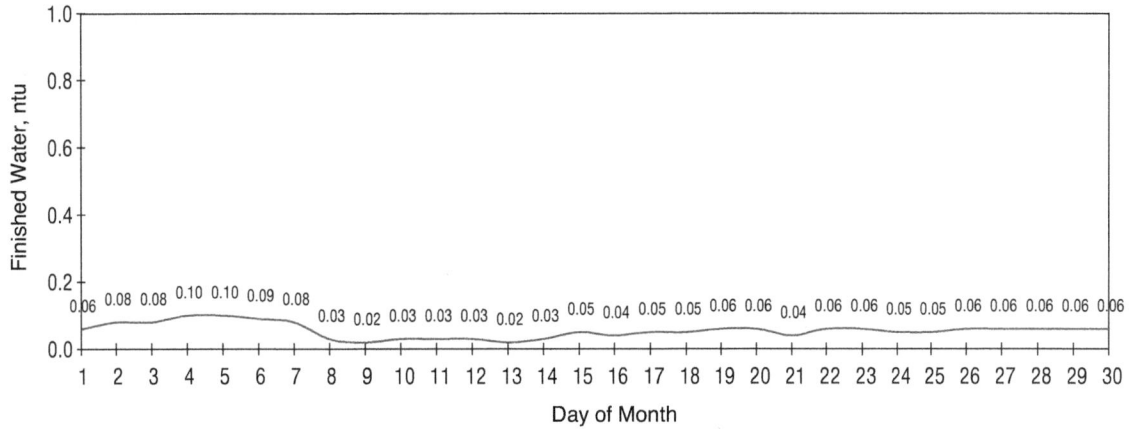

Source: Palm Beach County.

Figure 7.9-2   Turbidity after addition of ferric chloride, September 2006

The ferric chloride system at Palm Beach County Water Treatment Plant 8 is currently in service and has worked so well that Palm Beach County Water Utilities placed another system in Water Treatment Plant 2, another lime softening, ozone, filtration water treatment plant. Similar results have been obtained at WTP 2. The cost of this system is minimal compared to the benefits of reduction in other treatment chemicals and maintenance costs associated with the lime softening process.

This page intentionally blank.

Appendix: Examples of Standard Operating Procedures (SOPs)

This page intentionally blank.

Bull Creek SWTP
Coagulation–Flocculation Jar Test

1. Scope

1.1. This procedure covers a general procedure for the evaluation of a treatment to reduce dissolved, suspended, colloidal, and nonsettleable matter from water by chemical coagulation–flocculation, followed by gravity settling and filtration. This procedure may be used to evaluate color and turbidity removal along with residual aluminum content remaining after completion.

2. Summary

2.1. Coagulation–flocculation test is carried out to determine the chemicals, dosages, and conditions to achieve optimum results. The primary variables to be investigated are as follows:

    (A) Type chemicals added.
    (B) Chemical dosages added.
    (C) pH
    (D) Order of additions of chemicals and mixing conditions
    (E) Temperature.
    (F) Aluminum residuals in samples after coagulation.

3. Interferences

3.1. There are some possible interferences that may make the determination of optimum jar test conditions difficult. These include the following:

    (A)   Temperature changes during test. Thermal or convection currents may occur interfering with the settling of coagulated particles. (A waterbath should be used to reduce this interference.)
    (B)   Gas release during test. Flotation of coagulated floc may occur due to gas bubble formation caused by mechanical agitation, temperature increase or chemical reaction.

4. Reagents

4.1. Purity of Reagents—Plant grade chemicals should be used in all tests.
4.2. Raw Water—2 liter samples should be used in each beaker.

4.3. Jar test chemicals—Use graduated cylinders for water in the chemical make up below.

<u>Coagulants:</u>
(A) Alum—Add 6.2 ml's of raw plant alum into 200 ml's of distilled water. 1 ml of this solution will be 10 mg/l in a 2 liter raw water sample. Make a new batch every day used.
(B) Liquid Sodium Aluminate—Add 8.6 ml's of raw liquid sodium aluminate into 200 ml's of distilled water. 1 ml of this solution will be 10 mg/l in a 2 liter raw water sample. Make a new batch every day used.
(C) Dry Sodium Aluminate—Add 4.2 grams of dry sodium aluminate into 200 ml's of distilled water. 1 ml of this solution will be 10 mg/l in a 2 liter raw water sample. Make a new batch every day used.

<u>Coagulants Aids:</u>
(A) Liquid Polymer—Add .102 grams of raw liquid polymer into 500 ml's of distilled water. 1 ml of this solution will be .1 mg/l in a 2 liter raw water sample. Make a new batch every 2 weeks.
(B) Dry Polymer—Add .102 grams of dry polymer into 500 ml's of distilled water. 1 ml of this solution will be .1 mg/l in a 2 liter raw water sample. Make a new batch every 2 weeks.

<u>Taste & Odor Control :</u>
(A) Powder Activated Carbon—Add 1 gram of powder activated carbon into 500 ml's of distilled water. 1 ml of this solution will be 1 mg/l in a 2 liter raw water sample. Make a new batch every month.

(B) Potassium Permanganate—Add .206 grams of potassium permanganate to 500 ml's of distilled water. 1 ml of this solution will be 0.2 mg/l in a 2 liter raw water sample. Make a new batch every month.

<u>pH Adjustments</u>
(A) (Caustic Soda) Sodium Hydroxide—Add 1.3 ml of 50% plant caustic soda into 500 ml's of distilled water. 1 ml of this solution will be 1 mg/l in a 2 liter raw water sample. Make a new batch every month.
(B) Hydrated Lime—Add 1.01 grams of hydrated lime into 600 ml's of distilled water. 1 ml of this solution will be 1 mg/l in a 2 liter raw water sample. Make a new batch every month.

5. Procedures

5.1. Refer to Section 4.3 to determine the correct amounts of all chemicals to be used in the jar test. Measure out all proper dosages into the plastic syringes. 1 ml is equal to 1 cc on the syringe.

5.2. Calculate proper mix times for the 3 RPM mixing speeds. For the "flash mix block" on the jar test form; Take the raw water flow on the highest side of the plant and convert the flow to GPM then divide by 16,830. The answer will be the mixing time at 150 RPM's. Next take your same GPM reading on the highest side of the plant and divide it by 13,600. This answer will be the minutes of mixing at 50 RPM's. Next take the same highest raw flow in GPM and divide it by 3,141. This answer will be minutes of mixing at 25 RPM's. Insert answers on sheet.

Example: Total Raw Flow = 13.5 MGD
           East Flow = 7.0 MGD
           West Flow = 6.5 MGD

$$\frac{7,000,000}{1440} = 4,861 \text{ GPM}$$

Flash Mix = $\dfrac{4,861}{1440}$ = .29 minutes at 150 RPM

1st Stage = $\dfrac{4,861}{13,600}$ = .36 minutes at 50 RMP

2nd Stage = $\dfrac{4,861}{3,141}$ = 1.5 minutes at 25 RPM

JAR TEST DATA SHEET

	FLASH MIX	SLOW MIX 1ST STAGE	SLOW MIX 2ND STAGE	RAW WATER AMOUNT TESTED Liters	COLOR	TURB.NTU	ALKAL. mg/LCaCO₃	TEMP °C	pH	TDC mg/L	TSS mg/L	TDS mg/L	HARD mg/L CaCO₃
MINUTES	2	3	10	1	135	5.8	19	21.7	6	12		60	18
RPM	150	50	25	2									

Raw Water Characteristics

DATE / ANALYST / OBJECTIVE: Find correct alum dosage

JAR TEST FLASH MIX (MIXING TIME IN MINUTES) + FLOW INTO PLANT FLASH MIX (GPM) / 16830 GALLONS
1ST STAGE MIXING TIME IN MINUTES = FLOW INTO PLANT FLASH MIX (GPM) / 15600 GALLONS
2ND STAGE MIXING TIME IN MINUTES = FLOW INTO PLANT FLASH MIX (GPM) / 3,141 GALLONS

JAR NO.	PAC	Alum	Polymer	MINUTES to form first floc	DESCRIPTION OF FLOC	SETTLING RATE cm/sec	COLOR	TURB. NTU	ALKAL. mg/L CaCO₃	pH	Temp °C	Al³ RES mg/L	TOC mg/L
1	5	30	.2	Didn't form	NONE	N/A	--	--					
2	5	35	.2	30 sec	Size = 5.1 mm Normal floc	.2	30	1.5					
3	5	40	.2	30 sec	Size = 5.1 mm Normal floc	.2	15	0.23					
4	5	45	.2	30 sec	Size = 5.1 mm Normal floc	.2	12.5	0.2					
5	5	50	.2	30 sec	Size = 5.1 mm Normal floc	.2	13	0.27					
6	5	55	.2	30 sec	Size = 5.1 mm Normal floc	.2	13	0.2					

CHEMICALS USED mg/L In Order of Addition | FLOC CHARACTERISTICS | FILTRATE

ALAMEDA COUNTY WATER DISTRICT Standard Operating Procedure **WTP2**	NUMBER **34**	EFFECTIVE DATE: October 1, 1997 REVISED: **August 30, 2002**
SUBJECT: **Jar Test Procedure for WTP2, using the Pipette Method**		PAGE 1 OF 21

1.   Scope and Application

1.1   The jar test is used in the water treatment industry to simulate the treatment process. The test should be conducted whenever there is a significant change in water quality or as a tool for coagulant dosage optimization.

1.2   The jar test is performed in the laboratory to determine optimum dosage for chemical coagulants (usually ferric chloride or aluminum sulfate) and coagulant aids (cationic polymer) needed to achieve overall efficient plant operation.

1.3   The optimum coagulant dosage for a plant balances several factors:
- Settled water turbidity.
- Long filter runs.
- Organic removal.

Jar tests can be used to qualitatively evaluate the effects of different coagulant doses on these trends.

1.4   Jar tests are also used to compare and evaluate different coagulants and/or combinations of coagulants and coagulant aids.

1.5   This Pipette method saves time in preparing reagents. Also no dilution occurs when rinsing the reagent containers as in the conventional Beaker method.

1.6   Each test run may take 2-3 hours to complete.

2.   Apparatus and Materials

2.1   Phipps & Bird Stirrer model 7790-400 (figure 1).
2.2   Variable speed control (0-300 rpm).
2.3   6 @ 2 liter square jars, B-KER^2.
2.4   HACH turbidimeter 2100AN.
2.5   Vacuum pump.
2.6   Stopwatch.
2.7   6 @ Plastic bottles, 500 mL.
2.8   Thermometer.
2.9   Filter funnel and filter holder base (figure 5 - 6).
2.10   Membrane filter, Millipore, cat # HAWG 047 S3, filter type HA, 0.45 µm.
2.11   Vacuum flask (figure 5-6).
2.12   Septa, (12 one for each coagulant and coagulant aid used per jar) (figure 2.)
2.13   Illuminator, to provide light needed for a clear visual inspection.
2.14   Jar-test form (figure 10).
2.15   Graduated cylinder 20 and 250 mL.
2.16   1000 mL volumetric flasks (one for each stock solution).
2.17   Magnetic stir bars.
2.18   Magnetic stirrer.
2.19   Calculator.
2.20   Micro-pipette, two sizes: 10 to 100 µL (use yellow tips) and 100 to 1000 µL (use white tips) (figure 3.)

ALAMEDA COUNTY WATER DISTRICT Standard Operating Procedure WTP2	NUMBER 34	EFFECTIVE DATE: October 1, 1997 REVISED: **August 30, 2002**
SUBJECT: **Jar Test Procedure for WTP2, using the Pipette Method**		PAGE 2 OF 21

3.   **Reagents and Dilutions**

    3.1   **Ferric Chloride   FeCl$_3$**   coagulant

        Stock solution
- When to obtain a new sample:
- A new stock solution should be collected at least monthly or whenever a new delivery is received. The date of the most recent delivery can be found in the Chemical Delivery Binder found in the operators' console. **NOTE:** Ferric Chloride shipments are received often, so it is very likely that a week old stock solution is not from the most recent shipment and should be re-sampled so the date should always be checked.
- Obtaining a new sample:
- Dispose of previous month's storage in the laboratory sink.
- Rinse out the old bottle or a new one from the cabinet under the jar testing machine. Label the jar with the current date.
- If not an operator, ask the operator to obtain a ferric sample for you.

NOTE: Always notify the operator when taking a sample.
- Go to the ferric storage tank in the chemical bay with the empty bottle and a bucket for flushing the ferric line (use appropriate protective gear; gloves, goggles, raincoat).
- Slowly turn the sampling tap on the side of the tank and flush the line for 15 - 30 seconds into a bucket.
- Collect about 100 mL of ferric chloride

        Calculation of uL of ferric chloride to add to the septa for each 2 liter jar using an Eppendorf pipette
- The specific gravity of the Ferric solution as well as the percent activity of the solution is needed for the calculation. This information is found from the most recent Ferric delivery in the Chemical Delivery Binder in the operators' console.
- The calculation for the uL of ferric chloride to add for each mg/L of ferric chloride is as follows:

$$\frac{\text{Specific gravity}}{\text{\% activity}} = \left(\frac{1\ mg\ FeCl_3}{L\ water}\right)\left(\frac{L\ Ferric}{kg\ Ferric}\right)\left(\frac{kg}{10^6\ mg}\right)\left(\frac{kg\ Ferric}{kg\ FeCl_3}\right)\left(\frac{2\ L\ water}{bea\ ker}\right)\left(\frac{10^6\ uL}{L}\right)$$

        Which simplifies to :

$$\frac{2}{S.G * \%\ active} = \text{\# of } \mu\text{L Ferric solution to add per mg/L FeCl}_3 \text{ dose.}$$

ALAMEDA COUNTY WATER DISTRICT Standard Operating Procedure WTP2	NUMBER 34	EFFECTIVE DATE: October 1, 1997 REVISED: **August 30, 2002**
SUBJECT: **Jar Test Procedure for WTP2, using the** **Pipette Method**		PAGE 3 OF 21

EXAMPLE:

If the Ferric chloride most recent shipment is about 43% active and its specific gravity is 1.47:

$$\frac{2}{1.47*0.43} = \frac{2}{0.6321} = 3.16 \ \mu L \ \text{Ferric per mg/L FeCl}_3 \ \text{per jar}$$

In this case, the 3.16 µL is the amount to add to the 2 L beaker for each mg/ L of the ferric chloride desired in the jar test. So if you want to simulate a dose of 15 mg/L Ferric Chloride:

$$15 \frac{mg}{L} *3.16 \frac{uL}{mg/L} = 47.4 \ \mu L \ \text{Ferric Chloride should be placed on the septa.}$$

- **If you choose to vary the Ferric dose:**
- **If you want to optimize Ferric, you usually vary it by 2-5 parts per jar.**

EXAMPLE:

Using the same number for specific gravity and % activity as in the previous example:

	Jar 1	**Jar 2**	**Jar 3**	**Jar 4**	**Jar 5**	**Jar 6**
Desired ferric dose mg/L	15	20	25	30	35	40
µL of ferric to add to each septa	47.4	63.2	79.0	94.8	111	126

3.2   **Aluminum Sulfate (Alum)** $Al_2(SO_4)_3$ coagulant

Stock solution
- When to obtain a new sample:
- A new stock solution should be collected at least monthly or whenever a new delivery is received. The date of the most recent delivery can be found in the Chemical Delivery Binder found in the operators' console.
- Obtaining a new sample:
- Dispose of previous month's storage in the laboratory sink.
- Rinse out the old bottle or a new one from the cabinet under the jar testing machine. Label the jar with the current date.
- If not an operator, ask the operator to obtain an alum sample for you.
  NOTE: Always notify the operator when taking a sample.

ALAMEDA COUNTY WATER DISTRICT Standard Operating Procedure **WTP2**	NUMBER **34**	EFFECTIVE DATE: October 1, 1997 REVISED: **August 30, 2002**
SUBJECT: **Jar Test Procedure for WTP2, using the Pipette Method**		PAGE 4 OF 21

- Go to one of the alum feed pumps in the chemical bay with the empty bottle and a bucket for flushing the alum line (use appropriate protective gear; gloves, goggles, raincoat).
- Slowly turn the sampling tap on the feed pump and flush the line for 15 - 30 seconds into a bucket.

- Collect about 100 mL of alum

Calculation of uL of Alum to add to the septa for each 2 liter jar using an Eppendorf pipette

- This calculation is the same as the Ferric Chloride calculation and requires the same information: specific gravity and percent activity, which are obtained from the most recent Alum delivery in the Chemical Delivery Binder in the operators' console.
- Substituting Ferric with Alum, the equation simplifies to:

$$\frac{2}{S.G * \% \, active} \quad \text{\# of } \mu L \text{ Alum to add per mg/L Alum dose.}$$

EXAMPLE:

If the most recent shipment of Alum were 48% active and its specific gravity was 1.33:

$$\frac{2}{1.33 * 0.48} = 3.13 \text{ L Alum per mg/L alum per jar}$$

In this case, the 3.13 µL is the amount to add to the 2 L beaker for each mg/L of the alum desired. So if you want to simulate a dose of 40 mg/L alum:

$$40 \frac{mg}{L} * 3.13 \frac{uL}{mg/L} = 125 \, \mu L \text{ Alum should be placed on the septa.}$$

- **If you choose to vary the Alum Dose:**
- **If you want to optimize Alum, you can vary it by 2-5 parts per jar.**
- Each jar is calculated as shown for the Ferric.
  NOTE: Alum is rarely used in the plant. Ferric Chloride is the primary coagulant and Alum would be used in case of an emergency.

3.3   **Cationic Polymer (PE-C, polyDADMAC)   poly(dimethyl diallyl ammonium chloride) n($C_8H_{16}ClN$)**   coagulant aid

Stock solution
- When to obtain a new sample:

ALAMEDA COUNTY WATER DISTRICT Standard Operating Procedure WTP2	NUMBER 34	EFFECTIVE DATE: October 1, 1997 REVISED: **August 30, 2002**
SUBJECT: **Jar Test Procedure for WTP2, using the Pipette Method**		PAGE 5 OF 21

and

- A new stock solution should be collected at least monthly or whenever a new delivery is received. The date of the most recent delivery can be found in the Chemical Delivery Binder found in the operators' console. **NOTE:** The PE-C deliveries occur much less frequently than the Ferric therefore, a stock solution close to a month old may still be from the most recent delivery but the dates should always be checked.
- Obtaining a new sample:
- Dispose of previous month's storage in the laboratory sink.
- Rinse out the old bottle or a new one from the cabinet under the jar testing machine. Label the jar with the current date.
  If not an operator, ask the operator to obtain a PE-C sample for you.
  NOTE: Always notify the operator when taking a sample.
- Go to the PE-C storage tank in the chemical bay with the empty bottle and a bucket for flushing the PE-C line (use appropriate protective gear; gloves, goggles, raincoat).
- Slowly turn the sampling tap on the side of the tank and flush the line for 15 - 30 seconds into a bucket.
- Collect about 100 mL of PE-C

Intermediate Solution
- The cationic polymer used as a coagulant aid is very viscous, and should be diluted before using in a jar test.
- To make a **1:50 dilution** of the polymer (2%)
- use a 20 mL graduated cylinder to measure 20 mL of PE-C. Add to a 1000 mL volumetric flask. Rinse out the graduated cylinder with DI water into the flask to ensure you have removed all of the PE-C.
- Fill to the 1000 mL mark on the flask with deionized water. Mix the solution well using a stir bar if necessary.

Calculation of uL of PE-C working solution to add to the septa for each 2 liter jar using an Eppendorf pipette
- This calculation is similar to that of the Ferric but differs in the details.
- Although polymers from different manufacturers and batches contain varying amounts of active ingredients, for the purpose of calculations, we assume 100% active ingredient for the polyDADMAC product used at both treatment plants (as the operators assume when calculating plant doses.) As a result, the only information necessary from the Chemical Delivery binder is the specific gravity of the most recent shipment of PE-C

- Assuming

$$\left(\frac{1\,mg\,PE-C}{L\,water}\right)\left(\frac{L\,PE-C}{kg\,PE-C}\right)\left(\frac{kg}{10^6\,mg}\right)\left(\frac{1000\,L\,working\,solution}{20\,L\,PE-C}\right)\left(\frac{2\,L\,water}{bea\,\mathrm{ker}}\right)\left(\frac{10^6\,uL}{L}\right)$$

a 100 % active stock solution and taking into account the dilution to the working solution and specific gravity, the equation for PE-C is as follows:

Specific gravity		Working solution	
ALAMEDA COUNTY WATER DISTRICT Standard Operating Procedure **WTP2**	NUMBER **34**	EFFECTIVE DATE: October 1, 1997	
		REVISED: **August 30, 2002**	
SUBJECT: **Jar Test Procedure for WTP2, using the Pipette Method**		PAGE 6 OF 21	

$$\frac{100}{(s.g.)} =$$

Which simplifies to:

# of μL PE-C working solution to add per mg/L PE-C dose.

EXAMPLE:

If the most recent shipment of PE-C has a specific gravity of 1.08:

$$\frac{100}{1.08} = 92.59 \text{ L PE-C working solution per mg/L PE-C per jar}$$

In this case, the 92.59 μL is the amount to add to the 2 L beaker for each mg/L of the PE-C desired. So if you want to simulate a dose of 1.3 mg/L PE-C:

$$1.3\frac{mg}{L} * 92.59 \frac{uL}{mg/L} =$$ 120 μL PE-C working solution should be placed on the septa.

- **If you choose to vary the PE-C dose:**
- **If you want to optimize PE-C, you usually vary it by about 0.2 parts between jars.**

EXAMPLE:

Using the same number for specific gravity as in the previous example:

	Jar 1	Jar 2	Jar 3	Jar 4	Jar 5	Jar 6
Desired PE-C dose mg/L	0.9	1.1	1.3	1.5	1.7	1.9
μL of PE-C working solution to add to each septa	83.3	102	120	139	157	176

4.   **How to Use The Eppendorf Micro-Pipette** (figure 3)

4.1   **Size of Micropipette**
- Yellow top micropipette: This pipette is to measure over a range from 10 - 100 μL and is accurate to one decimal place. The clear tips in the yellow

boxes should always be used with this pipette.
- Blue Top micropipette: This pipette is for measuring amounts ranging from 100 - 1000 µL and is accurate to the ones place. The larger, clear tips in the blue boxes should always be used with this pipette.

ALAMEDA COUNTY WATER DISTRICT Standard Operating Procedure WTP2	NUMBER 34	EFFECTIVE DATE: October 1, 1997 REVISED: August 30, 2002
SUBJECT: **Jar Test Procedure for WTP2, using the Pipette Method**		PAGE 7 OF 21

4.2 Volume Setting

- Press the gray locking button found on the left side of the volume window down and hold.
- Set the volume by turning the control button.
  NOTE: Always set higher volumes first. When changing the volume from a lower to a higher setting, turn the knob past the desired volume and then back again.
- Release the locking button. The set volume is now secured against inadvertent adjustment.

4.3 Filling

- Attach pipette tip securely.
- Press the control button down to first the stop and hold.
- Hold pipette vertical and immerse the tip approximately 3 mm into the liquid.
  NOTE: Do not place a micro-pipette into liquid without a tip in place and do

  not immerse it further than the removable tip.
- Let the control button glide back slowly and smoothly.
- Slide the tip out of the liquid along the inside wall of the vessel, wiping off any droplets that remain on the outside of the tip.

4.4 Dispensing

- Hold tip upright and press the control button slowly down to the first stop to empty the tip onto the septa.
- Press the control button to the second stop to empty the tip completely.
- Let the control button glide back.

4.5 Removing Tip

- To remove a tip, hold the control button all the way to the third stop where it will release the tip. Throw the used tip in the trash.
- Replace the tip when you change between chemicals or if it becomes contaminated. Always throw away the tips after each jar test.

5.   Vacuum Pump / Filter Apparatus Operation (figure 5-6)

5.1 Set up the filter apparatus

- Open a 0.45 um Millipore HA membrane filter and carefully place it grid side up on the filter holder base that fits inside the flask.
- Place the magnetic filter funnel on top of the filter.

5.2 Filter Operation

- Turn on the vacuum pump and add a few drops of deionized water to just soak the filter paper. Wait for the vacuum to stabilize around 25 Hg.
- Using a graduated cylinder, measure out 200 mL of the water you are testing.

ALAMEDA COUNTY WATER DISTRICT Standard Operating Procedure WTP2	NUMBER 34	EFFECTIVE DATE: October 1, 1997 REVISED: **August 30, 2002**
SUBJECT: **Jar Test Procedure for WTP2, using the Pipette Method**		PAGE 8 OF 21

as

- Pour the 200 mL of test water into the filtration unit. Start the stopwatch soon as the first drop of the water has been poured into the filtration unit. Stop the watch when the first dry spot appears on the filter paper. Record the time, in seconds, as the filter time.
- **Blank:** This process should always first be done with deionized water as a reference. Ideally the DI water should filter in 20-30 seconds. If necessary, the vacuum pressure should be adjusted to achieve this DI water time. The **Filter Index** = sample time (seconds) / DI water time (seconds).
- During the jar test, this filter process is done several times on the settled water from the test. The blank must be done with the DI water and the applied water and finished water from the sample taps in the laboratory are also tested.

6.   **Special Requirements**

6.1   This procedure must be modified every time it is used to reflect conditions in the plant. Before beginning, record as much information on the jar test form as possible. Information such as raw water quality, number of flocculation and sedimentation basins, plant flow, plant dosages, and specifications for the most recent chemical deliveries will change with each test and are necessary to ensure that the jar test reflects the current plant conditions. Figure 10 contains a sample jar test form as well as where you can obtain all of the necessary data.

6.2   The doses to be tested will vary every time you do a jar test. Generally, routine jar tests are done to determine the best coagulant dose to be used in the plant. Ask the operator in charge his/her opinion about the current chemical dose. Generally, only one chemical is varied at a time and
two complete jar tests are run. Ask the operator which chemical to vary first and have him/her choose 6 doses such that the first beaker is under-dosed and the last is over-dosed. The middle beaker should contain the current chemical dose that the plant is using.
EXAMPLE: If the current plant doses are 19 mg/L Ferric and 1.2 mg/L PE-C, the first set of jars might test ferric doses of 10, 13, 16, 19, 22, and 25 mg/L with a constant PE-C dose at 1.2 mg/L. Once the test is run and an optimum ferric dose is determined, run the second test to determine the optimum PE-C dose. For example, if the optimum ferric dose was found to be 16 mg/L, the next test may keep the ferric dose constant at 16 mg/L and vary the PE-C doses of 0.6, 0.8, 1.0, 1.2, 1.4, and 1.6 mg/L.
NOTE: The doses that are obtained from the SCADA screens are not always accurate. If the operator has done a catch recently, use the calculated catch doses to decide on the doses to simulate in the jar test.

7.   **Procedure**

7.1   Ask an operator to turn on an **ozonated water pump. Flush** the sample line for at least 10 minutes before filling the beakers.

ALAMEDA COUNTY WATER DISTRICT Standard Operating Procedure WTP2	NUMBER **34**	EFFECTIVE DATE: October 1, 1997 REVISED: **August 30, 2002**
SUBJECT: **Jar Test Procedure for WTP2, using the Pipette Method**		PAGE 9 OF 21

7.2 Complete the preliminary information on the jar test form. Record raw water quality data, reason for testing, plant doses, etc. on the jar test form. The data that is collected in the laboratory (such as RW quality and DI and AW filter times) can also be done during the flocculation time to save some time.

7.3 Calculate the **flocculation time** using the following formula and record the time in the appropriate spot on the jar test form.

$$floc\ time,\ min = \frac{number\ of\ floc\ basins}{flow\ rate,\ mgd} \times 122$$

7.4 Calculate the **settling time** using the following formula and record the time in the appropriate spot on the jar test form.

$$settling\ time,\ min = \frac{number\ of\ sed\ basins}{flow\ rate,\ mgd} \times 17.9$$

7.5 Set two timers such that the first one goes off after flocculation is complete (1 minute + flocculation time) and the second beeps at the end of the test (1 minute + flocculation time + settling time.)

7.6 Calculate the **mixing speed (rpm)** to simulate the floc basins. First find the Velocity Gradient using the appropriate Velocity Gradient vs. Plant Flow rate graph (figure 7 or figure 8) (it depends on the number of open tubes per cell.) Then use the Laboratory G Curve graph (figure 9) to

determine the mixing speed on the x-axis from the velocity gradient on the y-axis. Record all times and speeds on the correct portion of the jar test form.

EXAMPLE for 7.3 through 7.5:

Plant flow = 21 MGD, 6 floc basins, 4 sed basins, and 4 tubes are in use.

$$Floc\ Time = \frac{6}{21} * 122 = 34.8 \approx 35\ minutes$$

$$Settling\ Time = \frac{4}{21} * 17.9 = 3.4 \approx 3.5\ minutes$$

Velocity gradient: Use figure 7. Find the velocity by finding where the rate intersects with the appropriate curve. In this case, 21 MGD crosses the 6 basin curve at 17.5 ft/sec.

Speed: Use figure 9. Where the velocity gradient (y-axis) is 17.5 ft/sec, this crosses the solid line at an impeller speed (x-axis) slightly above 20 rpm. We will use 20 rpm for simplicity.

ALAMEDA COUNTY WATER DISTRICT Standard Operating Procedure WTP2	NUMBER 34	EFFECTIVE DATE: October 1, 1997 REVISED: **August 30, 2002**
SUBJECT: **Jar Test Procedure for WTP2, using the Pipette Method**		PAGE 10 OF 21

On jar test form:

Time and Speed Sequence			
	Start Time	**Duration**	**Speed**
Rapid Mix	0 sec	30 sec	300 rpm
Slow Mix	30 sec	30 sec	200 rpm
Flocculation	1 min	35 min	20 rpm
Settling	36 min	3.5 min	0 rpm

End _39.5 min_

Set the timer for 36 minutes and 39.5 minutes.

7.7 **Decide on the doses for the chemical and calculate the amount of chemicals to be added to each jar** to simulate those doses (Section 3; most likely the chemicals added will be only ferric chloride (3.1) and the cationic polymer (3.3).)

7.8 Using the Eppendorf pipette as described in section 4, **dispense the calculated amount of ferric & PE-C onto the septa** (12 total septa) (figure 2.)

7.9 Fill the 6 square jar test containers to the 2 liter mark with **ozonated water** from the laboratory sink. Collect all 6 samples within a few minutes to avoid variability in the test results. Hooking a longer hose onto the sample tap makes this easier.

7.10 **Place the jars onto the Phipps and bird Stirrer, lower and tighten the stir paddles into the jars, and turn on the light on the apparatus** (figure 1.) Measure the temperature in one of the jars and record on the jar test form.

7.11 Turn on the mixer to **300 rpm**, make sure that the jars are all placed correctly such that the stir paddles are not hitting the jars and they are centered so that the septa will fall into the jars properly.

7.12 **Crank the septa bar such that the septa fall into the jar containers and start the timer as you crank the bar.**

7.13 At **30 seconds**. reduce the mixing speed to **200 rpm**.

7.14 At **1 minute** reduce the mixing speed to the **calculated flocculation speed** (section 7/6.)

7.15 **At the first timer (1 minute + flocculation time) turn off the mixer and allow the jars to settle.**

7.16 **Place the half liter square plastic bottles labeled 1 through 6 upright with the sample taps** (located on the side of the jars) inside the plastic bottles so that when the settling time is up, you can open all the taps at once and collect all samples.

7.17 **Important parameters to observe** during the jar test and record on the jar test form
- Relative speed for the first floc formation in each jar (first=1 and last=6).
- Relative floc size between the jars (VS=very small, S=small, M=medium, L=large, VL=very large).
- Relative settling rate between the jars (fastest=1 and slowest=6).
- Relative clarity of the water between floc particles between the jars (best=1 and worst=6).

  Record all observations and significant differences between jars on the bottom of the jar test form (flocculation example, figure 4.)

  NOTE: It is a good idea to have an operator, a senior operator, or a supervisor observe the jars with you to determine which jars look the best.

ALAMEDA COUNTY WATER DISTRICT Standard Operating Procedure WTP2	NUMBER 34	EFFECTIVE DATE: October 1, 1997 REVISED: **August 30, 2002**
SUBJECT: **Jar Test Procedure for WTP2, using the Pipette Method**		PAGE 11 OF 21

7.18   Immediately **after settling**, after the second timer is complete (1 minute + flocculation time + settling time), **collect approximately 500 mL of each jar sample** into the half liter square plastic bottles labeled 1 through 6. It is critical that the samples be collected from the sample tap,
located in the side of the jar test containers, all within a few minutes to avoid variability in the test results. If you have set up the plastic bottles as previously explained (section 7.16) you can open all the sample taps at once and simultaneously collect the samples from all jars.

7.19   **Analyze samples for turbidity** and record them as settled turbidity on the jar test form.

7.20   Determine the filtration time for deionized water, the applied water and filter effluent from the laboratory sink using the vacuum pump and the filter apparatus (section 5.) (This step can also be done earlier, while the jars are mixing to speed up the process.)

7.21   **Determine the filtration time for each jar sample** and record on the jar test form as filtered water filter time and calculate the filter index (section 5.)

7.22   **Analyze the filter effluent (filtrate) for turbidity for each jar sample** and record on the jar test form.

7.23   Make sure you have recorded all other necessary information on the jar test form and done all of the necessary laboratory analysis.

8.   **Analysis and Conclusions**

8.1   Upon completion of the test, go over the results and come to a conclusion by choosing the most optimal coagulant and doses.

8.2   Confer with the operator and discuss each aspect of the test results together.

- **Floc Characteristic:** The ideal floc particle will be the 2nd or 3nd rank in speed of forming. The flock will be large but no so large that it sinks to the bottom of the beaker during the flocculation stage of the test. It should hold together and not break apart as it is stirred. The rank for settling should be 2nd or 3nd. If the floc settles too quickly there is a chance that settling will occur in the floc basins before it reaches the sedimentation basins.
- **Settled Water:** The operators want to see very clear water between the floc particles as they are forming. The clearer the water, the cleaner the end product will be. The optimal beaker will be rank 1. The settled turbidity should coincide with the amount of chemical added. It will most likely increase as the dose decreases.
- **Filtered Water:** The optimal filtrate water will be the one that filters the fastest and produces the cleanest filtrate turbidity. It is common to have all the jars produce very similar filtrate turbidity, but the distinguishing factor will be the filter time.

ALAMEDA COUNTY WATER DISTRICT Standard Operating Procedure **WTP2**	NUMBER **34**	EFFECTIVE DATE: October 1, 1997 REVISED: **August 30, 2002**
SUBJECT: **Jar Test Procedure for WTP2, using the Pipette Method**		PAGE 12 OF 21

8.3   In the conclusion and comments on the jar test form, write a few sentences to summarize the test results. Note the jar number that produced the most optimal results and a brief description of why you and the operator selected it. Also discuss why the other jars failed, i.e., too much or too little coagulant or heavy floc.

8.4   The decision to change the plant's doses after the test is up to the operator. He/She may wish to examine the jar test form further.

8.5   Make sure a copy of the form is added to the jar test binder stored in the Laboratory.

9.   **Quality Control**

9.1   The stock solution of alum, ferric chloride and PE-C should not be kept longer than one month and should be re-sampled if there has been a new shipment since the sample was taken.
9.2   The prepared working solutions should be dated and kept no longer than 2 days.
9.3   Water samples should be collected immediately before the analysis.
9.4   Wash volumetric flasks immediately after test is finished.
9.5   Wash beakers immediately after test is finished.
9.6   Clean jars with 0.02 N sulfuric acid once a month.
9.7   Dispose of previous month's storage in the laboratory sink.
9.8   Never lay pipettes down with liquid in the tip.
9.9   Never pipette liquid without attaching a tip to the pipette or submerge the pipette past the tip.

10.   **References**

10.1   Introduction to Water Quality Analyses, Volume 4, American Water Works Association, 6666 W. Quincy Ave., Denver, CO 80235.
10.2   Water Quality, Principles and Practices of Water Supply Operations Series, Second Edition, 1995, American Water Works Association, 6666 W. Quincy Ave, Denver, CO 80235.
10.3   Water Treatment Handbook, Volume 1, Sixth Edition, 1991, Degremont, 11, rue Lavoisier-F 75384 Paris Cedex 08.
10.4   Operational Control of Coagulation and Filtration Processes, AWWA Manual M 37, 1992, American Water Works Association, 6666 W. Quincy Ave, Denver, CO 80235.

ALAMEDA COUNTY WATER DISTRICT Standard Operating Procedure WTP2	NUMBER 34	EFFECTIVE DATE: October 1, 1997
		REVISED: **August 30, 2002**
SUBJECT: **Jar Test Procedure for WTP2, using the Pipette Method**		PAGE 13 OF 21

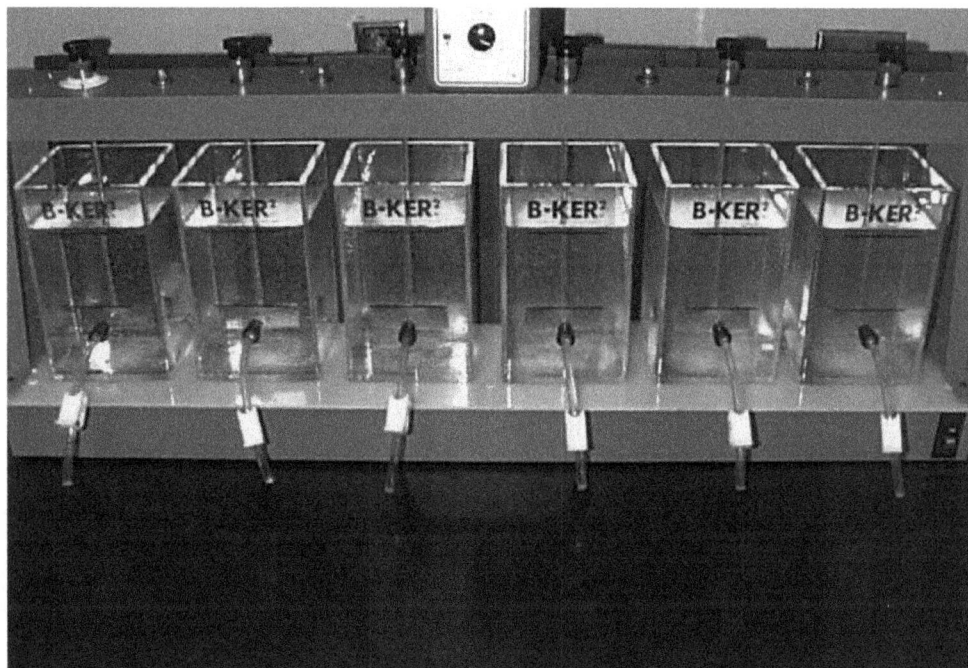

PHIPPS & BIRD STIRRER

FIGURE 1

ALAMEDA COUNTY WATER DISTRICT Standard Operating Procedure WTP2	NUMBER 34	EFFECTIVE DATE: October 1, 1997 REVISED: August 30, 2002
SUBJECT: **Jar Test Procedure for WTP2, using the Pipette Method**		PAGE 14 OF 21

THE SEPTA ARE LINING ON THE SEPTA BAR

CLOSE UP SECTION

FIGURE 2

ALAMEDA COUNTY WATER DISTRICT Standard Operating Procedure WTP2	NUMBER 34	EFFECTIVE DATE: October 1, 1997 REVISED: August 30, 2002
SUBJECT: **Jar Test Procedure for WTP2, using the Pipette Method**		PAGE 15 OF 21

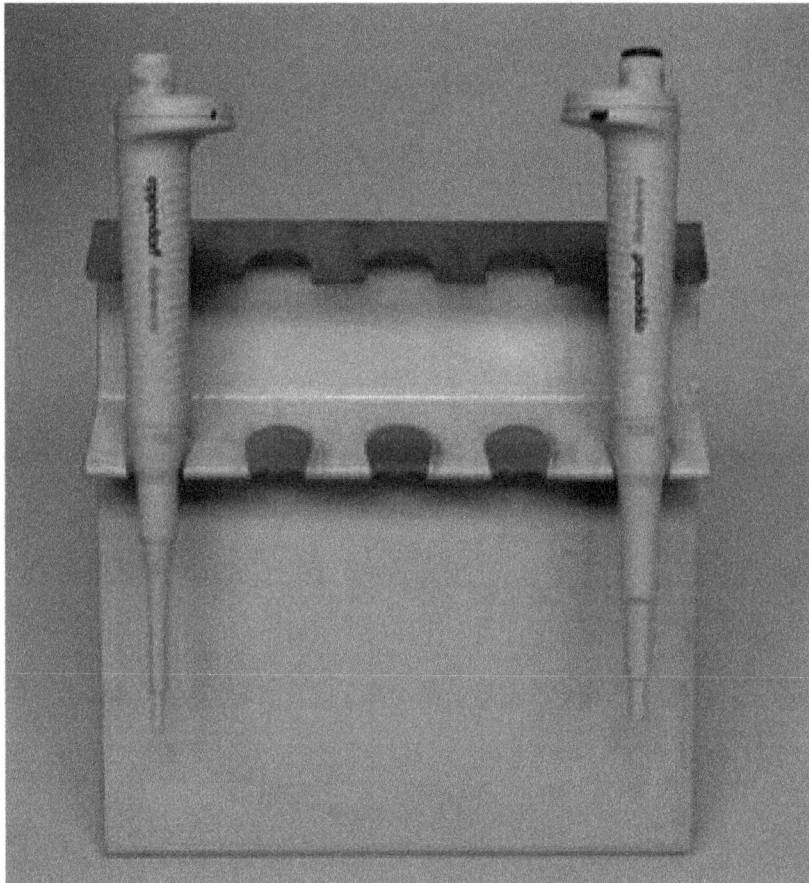

EPPENDORF MICRO-PIPETTE, SIZES, LEFT 10-100 μl, RIGHT 100-1000 μl

Figure 3

ALAMEDA COUNTY WATER DISTRICT Standard Operating Procedure WTP2	NUMBER 34	EFFECTIVE DATE: October 1, 1997 REVISED: August 30, 2002
SUBJECT: **Jar Test Procedure for WTP2, using the Pipette Method**		PAGE 16 OF 21

EXAMPLE OF FLOCCULATION

Figure 4

ALAMEDA COUNTY WATER DISTRICT Standard Operating Procedure WTP2	NUMBER 34	EFFECTIVE DATE: October 1, 1997 REVISED: August 30, 2002

SUBJECT: **Jar Test Procedure for WTP2, using the Pipette Method**	PAGE 17 OF 21

FILTRATION EQUIPMENT

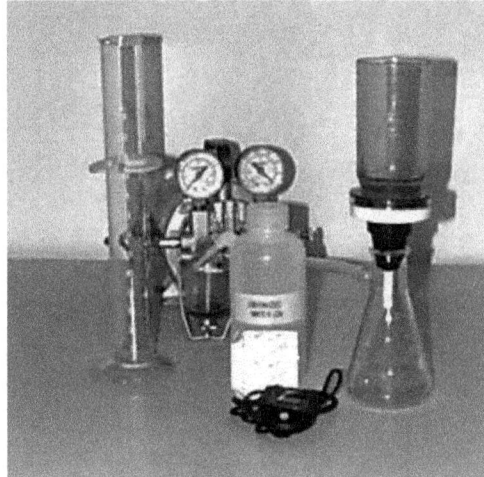

Figure 5

MEMBRANE FILTER   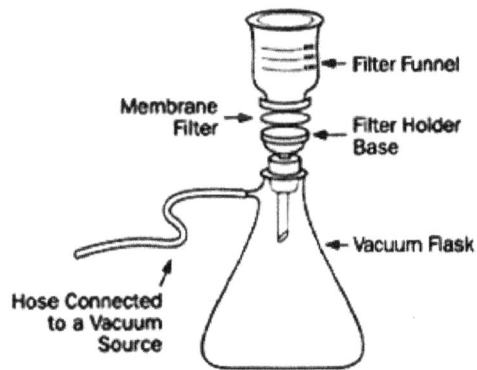   **EQUIPMENT**

Figure 6

ALAMEDA COUNTY WATER DISTRICT Standard Operating Procedure WTP2	NUMBER 34	EFFECTIVE DATE: October 1, 1997 REVISED: August 30, 2002

SUBJECT: **Jar Test Procedure for WTP2, using the Pipette Method**	PAGE 18 OF 21

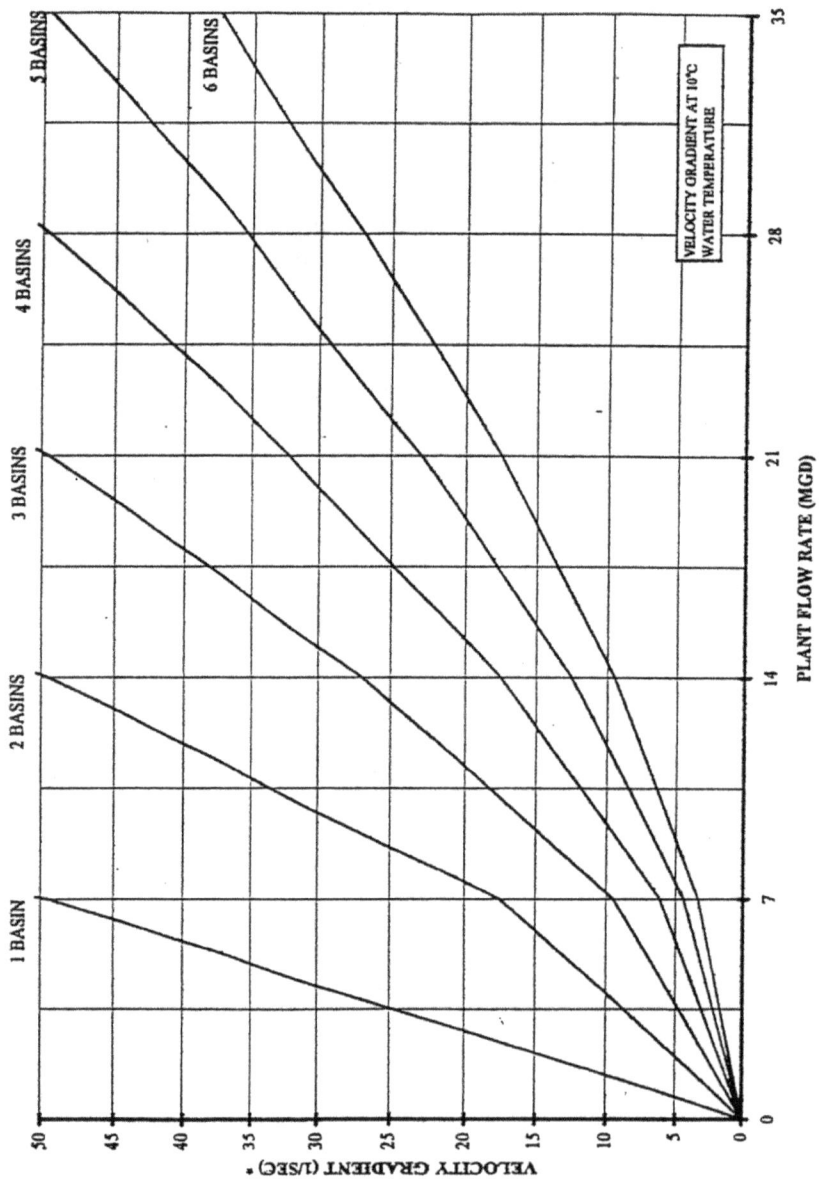

VELOCITY GRADIENT VS. PLANT FLOWRATE
WITH FOUR OPEN TUBES PER CELL
FIGURE 7

ALAMEDA COUNTY WATER DISTRICT Standard Operating Procedure **WTP2**	NUMBER **34**	EFFECTIVE DATE: October 1, 1997 REVISED: **August 30, 2002**

SUBJECT: **Jar Test Procedure for WTP2, using the Pipette Method**	PAGE 19 OF 21

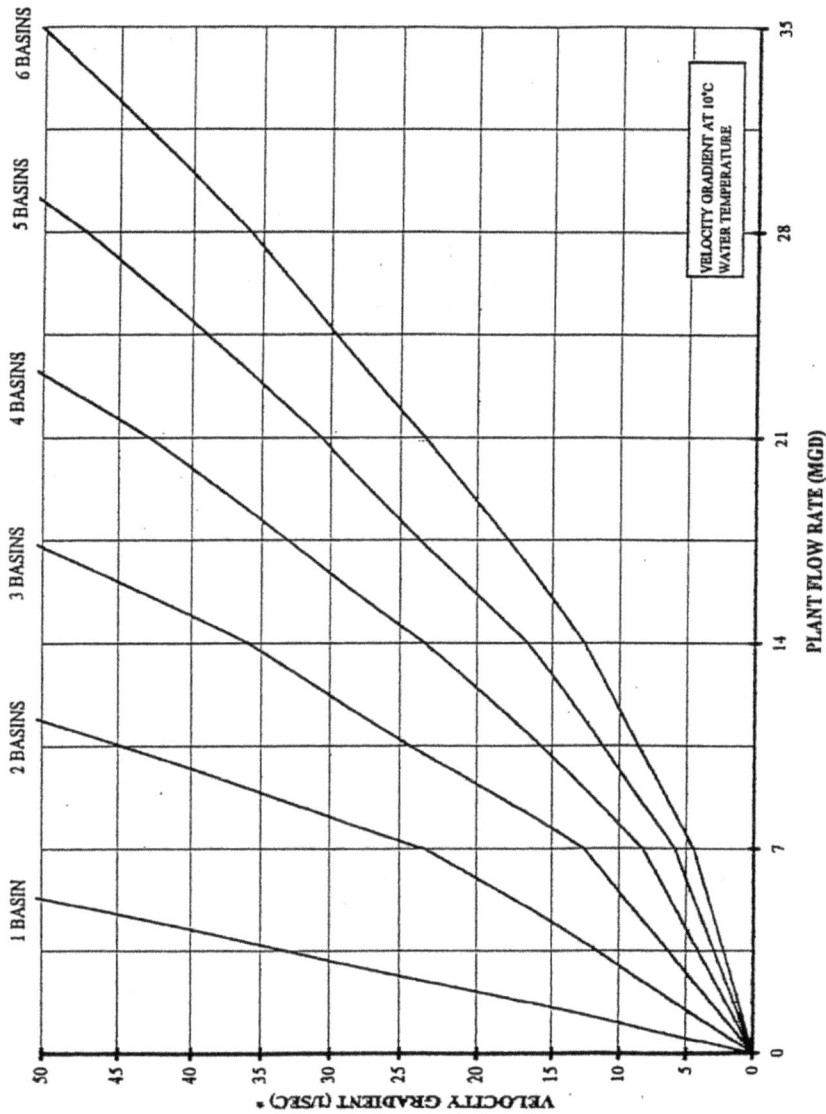

VELOCITY GRADIENT VS. PLANT FLOWRATE
WITH THREE OPEN TUBES PER CELL
FIGURE 8

ALAMEDA COUNTY WATER DISTRICT Standard Operating Procedure WTP2	NUMBER 34	EFFECTIVE DATE: October 1, 1997 REVISED: **August 30, 2002**
SUBJECT: **Jar Test Procedure for WTP2, using the Pipette Method**		PAGE 20 OF 21

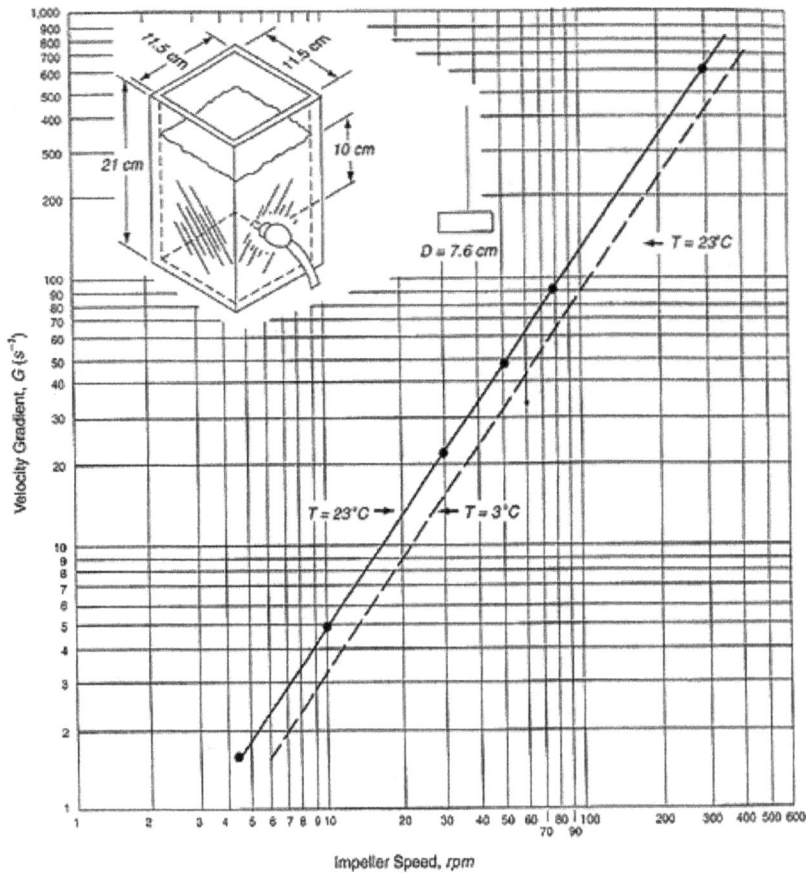

LABORATORY G CURVE

Figure 9

ALAMEDA COUNTY WATER DISTRICT Standard Operating Procedure WTP2	NUMBER 34	EFFECTIVE DATE: October 1, 1997 REVISED: August 30, 2002
SUBJECT: **Jar Test Procedure for WTP2, using the Pipette Method**		PAGE 21 OF 21

TREATMENT PLANTS JAR TEST FORM

Name:	Plant: TP1 ( ) TP2 ( )	Plant Rate: mgd	Time of Day:	Date:
Reason for Testing:		Method Used : (X)	Micro-pipette :	Beaker :

WTP Water Quality Data						Plant Process		*TP 2 Only*		Plant Dosages		Filter Index (FI) Test		
Raw Water	______% Delta	______% Del Valle				AW Turbidity	ntu	*# of Floc used*		Chlorine	mg/l	Filter Size/Type	0.45 um	Millipore-HA
Turbidity			Alkalinity			AW Filter Time	sec	*# of Seds used*		Ferric	mg/l	D. I. Water Time :		sec
pH			Spec Cond			AW Filter Index	(FI)	*# of Open Tubes*		PE-C	mg/l	Vacuum :		in Hg
Temp			Hardness			Filtered Water	ntu	*Ozone*	mg/l	Alum	mg/l	F I =	Sample Time / DI Time	

Chemicals Used (In Order of Addition)											Floc Characteristics			Settled Water		Filtered Water		
	Chlorine		Ferric		Alum		PE-C				Relative speed of forming	Relative Size	Relative Settling Rate	Relative Clarity Between Floc	Settled Turbidity	Filter Time	Filter Index	Filtrate Turbidity
Jar	mg/l	ul	mg/l	ul	mg/l	ul	mg/l	ul	mg/l	ul	(1-10)	(VS-VL)	(1-10)	(1-10)	(NTU)	(sec)	(FI #)	(NTU)

Ratings 1 = Best/Fastest 10 = Worst/Slowest S= Very Small S = Small M = Medium L = Large VL = Very Large

Time and Speed Sequence				Observations / Comments and Conclusions:
	Start Time	Duration	Speed	
Rapid Mix	sec	sec	rpm	
Slow Mix	sec	sec	rpm	
Flocculation	min	min	rpm	
Settling	min	min	rpm	

FIGURE 10

This page intentionally blank.

Index

AWWA Manuals

M1, *Principles of Water Rates, Fees, and Charges,* Fifth Edition, 2000, #30001PA

M2, *Instrumentation and Control,* Third Edition, 2001, #30002PA

M3, *Safety Practices for Water Utilities,* Sixth Edition, 2002, #30003PA

M4, *Water Fluoridation Principles and Practices,* Fifth Edition, 2004, #30004PA

M5, *Water Utility Management,* Second Edition, 2004, #30005PA

M6, *Water Meters—Selection, Installation, Testing, and Maintenance,* Second Edition, 1999, #30006PA

M7, *Problem Organisms in Water: Identification and Treatment,* Third Edition, 2004, #30007PA

M9, *Concrete Pressure Pipe,* Third Edition, 2008, #30009PA

M11, *Steel Pipe—A Guide for Design and Installation,* Fifth Edition, 2004, #30011PA

M12, *Simplified Procedures for Water Examination,* Fifth Edition, 2002, #30012PA

M14, *Recommended Practice for Backflow Prevention and Cross-Connection Control,* Third Edition, 2003, #30014PA

M17, *Installation, Field Testing, and Maintenance of Fire Hydrants,* Fourth Edition, 2006, #30017PA

M19, *Emergency Planning for Water Utility Management,* Fourth Edition, 2001, #30019PA 254 ductile -iron pipe and fitings

M20, *Water Chlorination / Chloramination Practices and Principles,* Second Edition, 2006, #30020PA

M21, *Groundwater,* Third Edition, 2003, #30021PA

M22, *Sizing Water Service Lines and Meters,* Second Edition, 2004, #30022PA

M23, *PVC Pipe—Design and Installation,* Second Edition, 2003, #30023PA

M24, *Dual Water Systems,* Third Edition, 2009, #30024PA

M25, *Flexible-Membrane Covers and Linings for Potable-Water Reservoirs,* Third Edition, 2000, #30025PA

M27, *External Corrosion—Introduction to Chemistry and Control,* Second Edition, 2004, #30027PA

M28, *Rehabilitation of Water Mains,* Second Edition, 2001, #30028PA

M29, *Fundamentals of Water Utility Capital Financing,* Third Edition, 2008, #30029PA

M30, *Precoat Filtration,* Second Edition, 1995, #30030PA

M31, *Distribution System Requirements for Fire Protection,* Fourth Edition, 2008, #30031PA

M32, *Distribution Network Analysis for Water Utilities,* Second Edition, 2005, #30032PA

M33, *Flowmeters in Water Supply,* Second Edition, 2006, #30033PA

M36, *Water Audits and Loss Control Programs,* Third Edition, 2009, #30036PA

M37, *Operational Control of Coagulation and Filtration Processes,* Third Edition, 2011, #30037PA

M38, *Electrodialysis and Electrodialysis Reversal,* First Edition, 1995, #30038PA

M41, *Ductile-Iron Pipe and Fittings,* Third Edition, 2009, #30041PA

M42, *Steel Water-Storage Tanks,* First Edition, 1998, #30042PA

M44, *Distribution Valves: Selection, Installation, Field Testing, and Maintenance,* Second Edition, 2006, #30044PA

M45, *Fiberglass Pipe Design,* Second Edition, 2005, #30045PA

M46, *Reverse Osmosis and Nanofiltration,* Second Edition, 2007, #30046PA

M47, *Capital Project Delivery,* Second Edition, 2010, #30047PA

M48, *Waterborne Pathogens,* Second Edition, 2006, #30048PA

M49, *Butterfly Valves: Torque, Head Loss, and Cavitation Analysis,* First Edition, 2001, #30049PA

M50, *Water Resources Planning,* Second Edition, 2007, #30050PA

M51, *Air-Release, Air / Vacuum, and Combination Air Valves,* First Edition, 2001, #30051PA

M52, *Water Conservation Programs—A Planning Manual,* First Edition, 2006, #30052PA

M53, *Microfiltration and Ultrafiltration Membranes for Drinking Water,* First Edition, 2005, #30053PA

M54, *Developing Rates for Small Systems,* First Edition, 2004, #30054PA

M55, *PE Pipe—Design and Installation,* First Edition, 2006, #30055PA

M56,   *Fundamentals and Control of Nitrification in*
        Chloraminated Drinking Water Distribution
        Systems, First Edition, 2006, #30056PA
M57,   *Algae: Source to Treatment,* First Edition, 2010,
        #30057PA
M58,   *Internal Corrosion Control in Water*
        Distribution Systems, First Edition, 2011,
        #30058PA